BestMasters

Konstantin Huter

Unraveling PFAS-Microplastic Interactions

In-Depth Insights Gained Through Laboratory Experiments and Computational Modeling Approaches

Konstantin Huter
Institute of Analytical Chemistry
and Radiochemistry
University of Innsbruck
Innsbruck, Austria

ISSN 2625-3577　　　　　　　ISSN 2625-3615　(electronic)
BestMasters
ISBN 978-3-658-49858-0　　　ISBN 978-3-658-49859-7　(eBook)
https://doi.org/10.1007/978-3-658-49859-7

This work is part of a project PFAS-Trap (Project number C220002), funded by the Federal Ministry Republic of Austria, Climate Action, Environment, Energy, Mobility, Innovation and Technology.

This Springer Spektrum imprint is published by the registered company Springer Fachmedien Wiesbaden GmbH, part of Springer Nature.
The registered company address is: Abraham-Lincoln-Str. 46, 65189 Wiesbaden, Germany

For Peter

Arbeit schreiben ist eine Hürde, aber kein Hindernis.

– Peter Huter

Acknowledgment I would first like to express my sincere gratitude to Univ.-Prof. Mag. Dr. Christian Huck for nominating this thesis to the Springer BestMasters series and for giving me the opportunity to work in his group.

A special note of gratitude in memory of our esteemed colleague, Priv.-Doz. Dr. Rania Bakry, whose invaluable discussions and support greatly enriched my research. Her absence is deeply felt at the institute.

Immense gratitude goes to Assoz.-Prof. Dr. Matthias Rainer for not only including me in the institute but also for creating such a welcoming and collaborative atmosphere. This environment truly enriched my experience and made every day enjoyable. The exchanges with him were always very pleasant.

I would like to express my sincere gratitude to Assoz.-Prof. Dr. Thomas Hofer for his invaluable assistance with the computational calculations. Our meetings with him were not only highly informative and productive but also enjoyable and engaging. I would also like to thank Dr. Jan Back, from the MCI, for his generous support in conducting the measurements for the differential volume distribution.

I must also thank Dr. Christoph Kappacher and Dr. Johanna Freilinger for their patience and the spirited environment they maintained. Their support made the demanding work not only easier but also enjoyable. Johanna's management of our LC-MS/MS was challenging, but she handled it exceptionally well and I did learn to never give up (on your instruments).

Thanks to Peter Rutzinger, MSc. for putting up with the never-ending racket from the centrifugal mill and to Jovan Badzoka, MSc. for all the practical tips and tricks in the lab.

Additionally, I want to thank everyone at the Institute of Analytical Chemistry for warmly welcoming me. The coffee breaks and lunches were a true highlight. I have genuinely enjoyed working with all of you and I am excited to continue working together.

To my incredible friends, thank you for the laughter and for making university life so much brighter. I truly could not have done it without you.

To my girlfriend Nadine, whose support has been invaluable throughout both my bachelor's thesis and master's thesis. A heartfelt thanks to my family, especially my father, Peter. His constant support and humor guide and inspire me to this day. I am sure he would be very proud.

Competing Interests The author has no competing interests to declare that are relevant to the content of this manuscript.

Abstract

This thesis investigates the adsorption interactions between per- and polyfluoroalkyl substances (PFAS) and a diverse set of synthetic polymers, highlighting their roles as persistent environmental contaminants. By integrating experimental techniques, notably liquid chromatography-tandem mass spectrometry (LC-MS/MS), with computational modeling via the MACE-OFF23 neural network potential and GFN2-xTB/GBSA, this study provides new insights into how PFAS interact with various polymers.

Eight PFAS in combination with 18 polymers were systematically examined under different pH conditions to understand how polymer composition and PFAS polarity influence adsorption efficiency. Here it was shown that polyamides (PAs) achieved up to 100% uptake (8 mg PFAS per g polymer), primarily due to hydrogen bonding and electrostatic interactions at amide linkages. In contrast, polyethylene (PE) exhibited minimal adsorption, showing that specific functional groups (especially amide or carbonyl units) are more important than overall hydrophobicity. Adjusting the pH of the solution had a profound impact on PFAS uptake in amide-rich polymers, underscoring the significance of electrostatic forces and hydrogen bonding under acidic to alkaline conditions. Computational calculations supported these findings, revealing stronger binding of PFAS, especially perfluorobutanoic acid (PFBA) to amide-bearing polymers (-126 kJ mol^{-1} for PA-PFBA) than to pure hydrocarbon based polymers (PE) (-39 kJ mol^{-1} for PE-PFBA).

These results demonstrate the strong influence of both polymer functionality and environmental factors on PFAS adsorption, explaining why polyamides and similar polymers are especially well-suited to PFAS remediation strategies.

Future investigations should explore more complex matrices that include salinity, competing ions, and varying environmental parameters, along with refined computational models to predict PFAS adsorption across a broader range of conditions. This combined experimental and theoretical approach clarifies the mechanisms of PFAS–polymer interactions and offers practical guidelines for selecting or engineering polymers to reduce PFAS contamination effectively.

Zusammenfassung

Diese Arbeit untersucht die Adsorptionsinteraktionen zwischen per- und polyfluorierten Alkylsubstanzen (PFAS) und einer vielfältigen Auswahl synthetischer Polymere, wobei deren Rolle als persistente Umweltkontaminanten hervorgehoben wird. Durch die Kombination experimenteller Methoden, insbesondere Flüssigchromatographie-Tandem-Massenspektrometrie (LC-MS/MS), mit computergestützten Modellen, wie dem MACE-OFF23-Neuronalen Netzwerkpotenzial und GFN2-xTB/GBSA, liefert diese Studie neue Erkenntnisse darüber, wie PFAS mit verschiedenen Polymeren interagieren.

Acht PFAS wurden in Kombination mit 18 Polymeren systematisch unter unterschiedlichen pH-Bedingungen untersucht, um zu verstehen, wie die Polymerzusammensetzung und die Polarität der PFAS die Adsorptionseffizienz beeinflussen. Dabei zeigte sich, dass Polyamide (PA) eine Aufnahme von bis zu 100% (8 mg PFAS pro g Polymer) erreichten, was vor allem auf Wasserstoffbrückenbindungen und elektrostatische Wechselwirkungen an den Amidbindungen zurückzuführen ist. Im Gegensatz dazu zeigte Polyethylen (PE) eine minimale Adsorption, was darauf hinweist, dass spezifische funktionelle Gruppen (insbesondere Amid- oder Carbonyleinheiten) wichtiger sind als die allgemeine Hydrophobie. Die Anpassung des pH-Werts der Lösung hatte einen erheblichen Einfluss auf die PFAS-Aufnahme in amidreichen Polymeren, was die Bedeutung elektrostatischer Kräfte und Wasserstoffbrückenbindungen unter sauren bis alkalischen Bedingungen unterstreicht. Computergestützte Berechnungen unterstützten diese Ergebnisse, indem sie eine stärkere Bindung von PFAS, speziell Perofluorobutansäure (PFBA), an amidhaltige Polymere (-126 kJ mol^{-1} für PA-PFBA) im Vergleich zu rein kohlenstoffbasierten Polymeren (PE) (-39 kJ mol^{-1} für PE-PFBA) zeigten.

Diese Ergebnisse verdeutlichen den starken Einfluss sowohl der Polymer-funktionalität als auch der Umweltfaktoren auf die PFAS-Adsorption und erklären, warum Polyamide und ähnliche Polymere besonders gut für PFAS-Entfernungsstrategien geeignet sind. Zukünftige Untersuchungen sollten komplexere Matrizes berücksichtigen, die Salinität, konkurrierende Ionen und variierende Umweltparameter einbeziehen, sowie verfeinerte Computermodelle entwickeln, um die PFAS-Adsorption unter einem breiteren Spektrum von Bedingungen vorherzusagen. Dieser kombinierte experimentelle und theoretische Ansatz klärt die Mechanismen der PFAS-Polymer-Interaktionen und bietet praktische Leitlinien zur Auswahl oder Entwicklung von Polymeren, um die PFAS-Belastung wirksam zu reduzieren.

Contents

Abbreviations

ABS	Acrylonitrile butadiene styrene
APCI	Atmospheric pressure chemical ionization
ASA	Acrylonitrile-styrene-acrylate copolymer
CA	Cellulose acetate
DFTB+	Density Functional Tight Binding Plus
EC	European Commission
EPA	Environmental Protection Agency
ESI	Electrospray ionization
EVAC	Ethylene-vinyl acetate copolymer
FA	Formic acid
GBSA	Generalized Born with surface area
Gen-X	Ammonium perfluoro (2-methyl-3-oxahexanoate)
GFN2-xTB	The second-generation extended tight-binding method
GO	Global optimization
HDPE	High-density polyethylene
HPLC	High performance liquid chromatography
IRIS	Integrated Risk Information System
IS	Internal standard
LC	Liquid chromatography
LC-MS/MS	Liquid chromatography-tandem mass spectrometry
LDPE	Low-density polyethylene
LLDPE	Linear low-density polyethylene
MACE-OFF23	A specific neural network potential used for global optimization

MeOH	Methanol
MP	Microplastic
MRM	Multiple reaction monitoring
MS	Mass spectrometry
NNP	Neural network potential
PA 10.10	Polyamide 10.10
PA 6	Polyamide 6
PA 6.6	Polyamide 6.6
PC	Polycarbonate
PE	Polyethylene
PET	Polyethylene terephthalate
PFAS	Per- and polyfluoroalkyl substances
PFBA	Perfluorobutanoic acid
PFBS	Perfluorobutanesulfonic acid
PFHxA	Perfluorohexanoic acid
PFNA	Perfluorononanoic acid
PFOS	Perfluorooctanesulfonic acid
PFOA	Perfluorooctanoic acid
PFPeA	Perfluoropentanoic acid
PLA	Polylactic acid
PMMA	Polymethylmethacrylate
POM	Polyoxymethylene
POPs	Persistent organic pollutants
PP	Polypropylene
PS	Polystyrene
PTFE	Polytetrafluoroethylene
PVC	Polyvinyl chloride
PVDF	Polyvinylidene fluoride
R^2	Coefficient of determination
REACH	Regulation concerning the Registration, Evaluation, Authorisation, and Restriction of Chemicals
R_t	Retention time
SVHC	Substances of Very High Concern
TPU	Thermoplastic polyurethane elastomers
UHPLC	Ultra high performance liquid chromatography

List of Figures

List of Tables

Introduction

1.1 Theoretical Background

1.1.1 Per- and Polyfluoroalkyl Substances (PFAS)

Per- and polyfluoroalkyl substances (PFAS) are a group of synthetic chemicals which contain aliphatic carbon atoms where hydrogen atoms have been replaced by fluorine atoms.[1–5] PFAS can include both, perfluoroalkyl compounds (all hydrogen atoms on the carbon chain are replaced by fluorine atoms) and polyfluoroalkyl compounds (at least one but not all hydrogen atoms are replaced by fluorine atoms).[1, 6, 7]

PFAS typically consist of a carbon chain (either fully or partially fluorinated) that is hydrophobic along a functional group that is hydrophilic which has adjustable functional groups like carboxylic- or sulfonic acids, as seen in Fig. 1.1.[8] The carbon-fluorine bond (C–F) exhibits the strongest bond in organic chemistry, with a bond dissociation energy of $485\,\mathrm{kJ\,mol^{-1}}$.[9, 10] This exceptional bonding strength is attributed to the high electronegativity of fluorine, which leads to a significant increase in bond energy. As a result, per- and polyfluoroalkyl substances (PFAS) are chemically and thermodynamically stable, rendering them resistant to degradation thus making them alarming for health effects. Consequently, PFAS are widely utilized in various industrial applications due to their exceptional stability and durability.[11, 12]

In addition to these classifications, PFAS are also commonly distinguished by the length of their fluorinated carbon chain, often referred to as long-chain versus short-chain PFAS.[13–17] Long-chain PFAS generally include substances with eight or more perfluorinated carbon atoms, such as perfluorooctanoic acid (PFOA) and perfluorooctanesulfonic acid (PFOS), whereas short-chain PFAS have fewer than eight perfluorinated carbon atoms (e.g., perfluorohexanoic acid [PFHxA] and perfluorobutanesulfonic acid [PFBS]). These differences in chain length can

K. Huter, *Unraveling PFAS-Microplastic Interactions*, BestMasters, https://doi.org/10.1007/978-3-658-49859-7_1

influence both environmental fate and toxicological profiles. Long-chain PFAS tend to be more persistent and bioaccumulative, remaining in organisms and the environment over extended periods. In contrast, short-chain PFAS, while still resistant to degradation, exhibit lower bioaccumulation potential and are more mobile in aquatic systems. Nevertheless, the growing use of short-chain PFAS as substitutes for their long-chain counterparts warrants close monitoring, since they can still pose significant environmental and health concerns.[17]

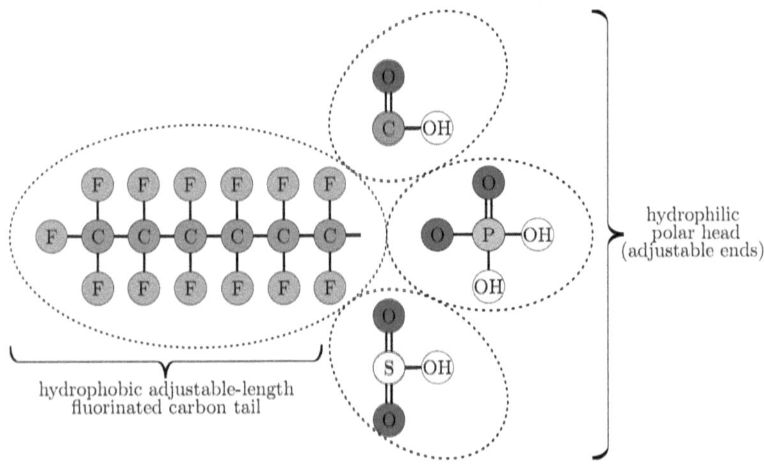

Fig. 1.1 An overview of the structure of PFAS in general adapted from literature[8]

To facilitate a comprehensive understanding of each PFAS discussed in this study, the subsequent sections will provide detailed chemical structures, concise explanations, relevant applications, and pertinent information.

1.1.1.1   Perfluorooctanoic Acid (PFOA)

Perfluorooctanoic acid (PFOA) is a synthetic perfluorinated compound with the chemical formula $C_8HF_{15}O_2$, introduced by 3M in the late 1940s. The molecular structure of PFOA is shown in Fig. 1.2. It is widely detected in human serum, with a median concentration of 4 $ng\,mL^{-1}$ in the United States.[18] The primary sources of exposure remain unclear, though wastewater treatment plants are considered significant contributors.[19, 20]

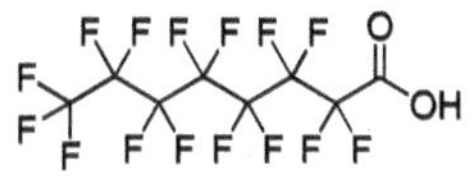

Fig. 1.2 Structure of perfluorooctanoic acid (PFOA)

PFOA has been utilized in a variety of industrial and consumer products, including Teflon™, Gore-Tex®, paper, food packaging, and clothing.[18–20] From 1951 to 2004, global production of PFOA was estimated to range between 3600 and 5700 tons.[21, 22] However, projections from 2005 to 2050 anticipate a reduction to between 480 and 950 tons, primarily due to regulatory actions such as the EU's ban on PFOA in 2020.[23] Additionally, 3M discontinued the production and use of PFOA in the United States in 2002. Since 2015, the Stewardship Program has engaged eight major companies in the PFAS industry to commit to reducing PFOA production.[24, 25]

Exposure to PFOA has been linked to various health effects, including cardiovascular disease and peripheral arterial diseases.[26] The largest health study on PFOA, involving over 69,000 individuals, revealed connections with kidney, ovarian, testicular, and prostate cancers, as well as non-Hodgkin lymphoma.[27–29]

Regulations concerning PFOA in drinking water vary globally. In the EU, a limit of $0.025\,\mathrm{mg\,kg^{-1}}$ for PFOA in products has been established.[23] In the United States, guideline levels for PFOA in drinking water range from 13 to 1000 $\mathrm{ng\,L^{-1}}$, varying by state.[30]

1.1.1.2 Perfluorobutanoic Acid (PFBA)

Perfluorobutanoic acid (PFBA) is a perfluoroalkyl carboxylic acid with the molecular formula $C_3F_7CO_2H$. The structure is shown in Fig. 1.3. As a breakdown product of longer-chain PFAS, PFBA is utilized across various industries, including food packaging, firefighting foams, and consumer products.[1, 31] Notably, PFBA has been detected in chicken eggs in China and in groundwater in India at concentrations of 9.2 $\mathrm{ng\,L^{-1}}$.[32, 33] The EPAs Integrated Risk Information System (IRIS) indicates that there is currently inadequate information to assess the carcinogenic potential of PFBA, reflecting a lack of conclusive data.[34] Although it exhibits lower bioaccumulation in human tissues compared to longer-chain PFAS, PFBA is still associated with significant health risks, including systemic toxicity and biological disruption.[35–38]

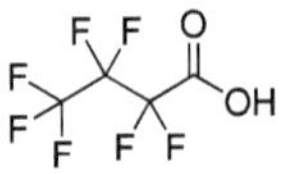

Fig. 1.3 Structure of perfluorobutanoic acid (PFBA)

1.1.1.3 Perfluorooctanesulfonic Acid (PFOS)

Perfluorooctanesulfonic acid (PFOS) is an eight-carbon fluorocarbon with the formula $C_8HF_{17}O_3S$, originally developed by the 3M Company. The chemical is depicted in Fig. 1.4 It was a key ingredient in Scotchgard, a stain repellent product by 3M.[39] The applications of PFOS are extensive, ranging from firefighting foams to coatings and various industrial uses.[1, 7, 11, 40, 41] The chemical was phased out of production by 3M in 2000, coinciding with a broader ban on many PFOS applications.[24, 25, 42]

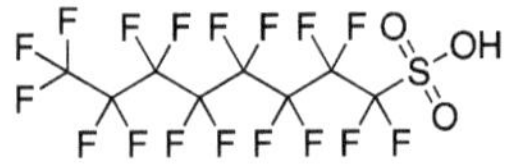

Fig. 1.4 Structure of perfluorooctanesulfonic acid (PFOS)

In 2006, an EC Directive (2006/122/ECOF) in Europe restricted the use of PFOS. In 2009, PFOS was added to Annex B of the Stockholm Convention on Persistent Organic Pollutants (POPs). In 2015, over 200 scientists endorsed the Madrid Statement, which called for international limits on the production and use of PFAS, particularly expressing concerns over the emergence of new short-chain perfluorinated compounds.[43]

PFOS exposure has been linked to a range of adverse health effects, including metabolic disorders, immune suppression, endocrine disruption, and developmental toxicity.[44–51] In the United States, regulatory limits for PFOS in drinking water range from 13 to 1000 ngL^{-1}, varying by state, similar to those for PFOA.[30] International regulations on PFOS mirror those applied to PFOA, demonstrating a global approach to managing its risks.

1.1.1.4 Ammonium Perfluoro(2-methyl-3-oxahexanoate)(Gen-X)

Gen-X is the trademark of Chemours for a synthetic, short-chain organofluorine PFAS with the formula $C_6H_4F_{11}NO_3$. The ether linked Gen-X structure is displayed in Fig. 1.5. It was developed by DuPont in 2009 as a replacement for PFOA and is utilized as a processing aid in the manufacture of various fluoropolymers, including polytetrafluoroethylene (PTFE). Its applications extend to products such as food packaging, paints, and firefighting foams.[52] Due to its relatively recent introduction, comprehensive data on Gen-X concentrations in drinking water and the environment are limited. Nonetheless, notable findings include concentrations exceeding 4000 $\mathrm{ng\,L^{-1}}$ detected in private drinking water wells near a manufacturing plant in North Carolina.[52, 53] In Europe and Asia, reported concentrations reach up to 108 $\mathrm{ng\,L^{-1}}$ in Germany, 91.5 $\mathrm{ng\,L^{-1}}$ in the Netherlands, and 3830 $\mathrm{ng\,L^{-1}}$ in China.[54] Additionally, a maximum concentration of 812 $\mathrm{ng\,L^{-1}}$ was recorded in the Rhine River in the Netherlands.[55] Notably, these samples were collected downstream from PFAS manufacturing plants. Moreover, similar to PFOA, Gen-X has been linked to a variety of health issues, underscoring the need for ongoing research and regulatory scrutiny.[56, 57]

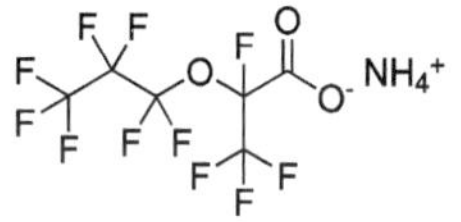

Fig. 1.5 Structure of ammonium perfluoro(2-methyl-3-oxahexanoate) (Gen-X)

1.1.1.5   Perfluoropentanoic Acid (PFPeA)

Perfluoropentanoic acid (PFPeA) is a short-chain perfluorocarboxylic acid with the formula $C_5HF_9O_2$.[58] See Fig. 1.6 for the PFPeA structure. It is believed that PFPeA forms as a byproduct of the degradation of longer-chain PFAS compounds such as PFOA.[59] Notably, PFPeA has been detected in a broad spectrum of matrices including shellfish and hot cocoa.[60] Furthermore, the presence of PFPeA in serum and urine of children indicates continuous exposure to this compound.[61]

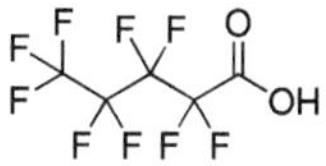

Fig. 1.6 Structure of perfluoropentanoic acid (PFPeA)

1.1.1.6 Perfluorobutanesulfonic Acid (PFBS)

Perfluorobutanesulfonic acid (PFBS) is a short-chain PFAS molecule with the formula $C_4HF_9O_3S$ and its molecular structure is shown in Fig. 1.7. Since 2003, it has been used by the 3M Company as a safer alternative to PFOS. Notably, PFBS has been detected in environmental samples, with concentrations reaching 10.2 $ng\,L^{-1}$ in river water and 4.9 $ng\,L^{-1}$ in groundwater.[33] These findings underline the persistent nature of even newer PFAS compounds in the environment. On January 16, 2020, PFBS and its salts were added to the REACH Regulation Candidate List of Substances of Very High Concern (SVHCs) due to their potential serious health effects, affirming their "equivalent level of concern" compared to other hazardous substances.[62]

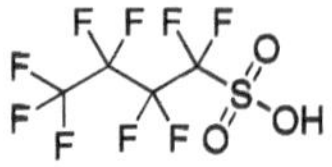

Fig. 1.7 Structure of perfluorobutanesulfonic acid (PFBS)

1.1.1.7 Perfluorohexanoic Acid (PFHxA)

Perfluorohexanoic acid (PFHxA) is a short-chain perfluoroalkyl acid with the formula $C_6HF_{11}O_2$. Its six carbon perfluoroalkyl chain is drawn in Fig. 1.8. Although it is used as a substitute for the more harmful long-chain PFAS compounds, PFHxA still presents considerable health and environmental risks. Recent studies have highlighted its potential to cause liver toxicity, developmental disruptions, neurotoxic effects, and significant ecological impacts.[37, 63–69]

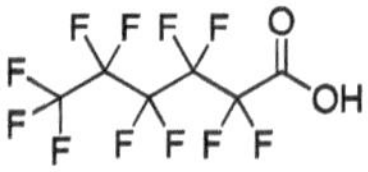

Fig. 1.8 Structure of perfluorohexanoic acid (PFHxA)

1.1.1.8 Perfluorononanoic Acid (PFNA)

Perfluorononanoic acid (PFNA) is a perfluorinated carboxylic acid with the formula $C_9HF_{17}O_2$. The nine carbon perfluoroalkyl carboxylate is shown in Fig. 1.9. It is recognized as a developmental toxicant and an immune system disruptor.[70] Notably, PFNA has been detected in human follicular fluid and in dolphin blood plasma, where concentrations exceeded 100 ppb.[71, 72]

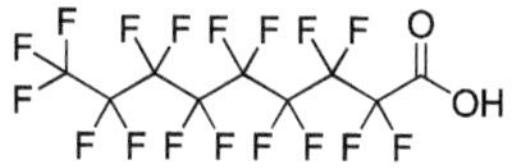

Fig. 1.9 Structure of perfluorononanoic acid (PFNA)

1.1.2   Microplastic (MP)

MPs are defined as plastic particles ranging in size from 1 μm to 5 mm.[73–76] These particles are further categorized into two types based on their origin. Primary MPs are intentionally manufactured small plastic particles, commonly found in products such as cosmetics, personal care items, and industrial abrasives.[76, 77] Secondary MPs, on the other hand, are the result of fragmentation and degradation of larger plastic debris caused by physical, chemical, and biological processes such as photo-oxidation by sunlight and mechanical abrasion, and they are found in both marine and land-based environments.[78]

The polymers analyzed in this study are depicted in Fig. 1.10.

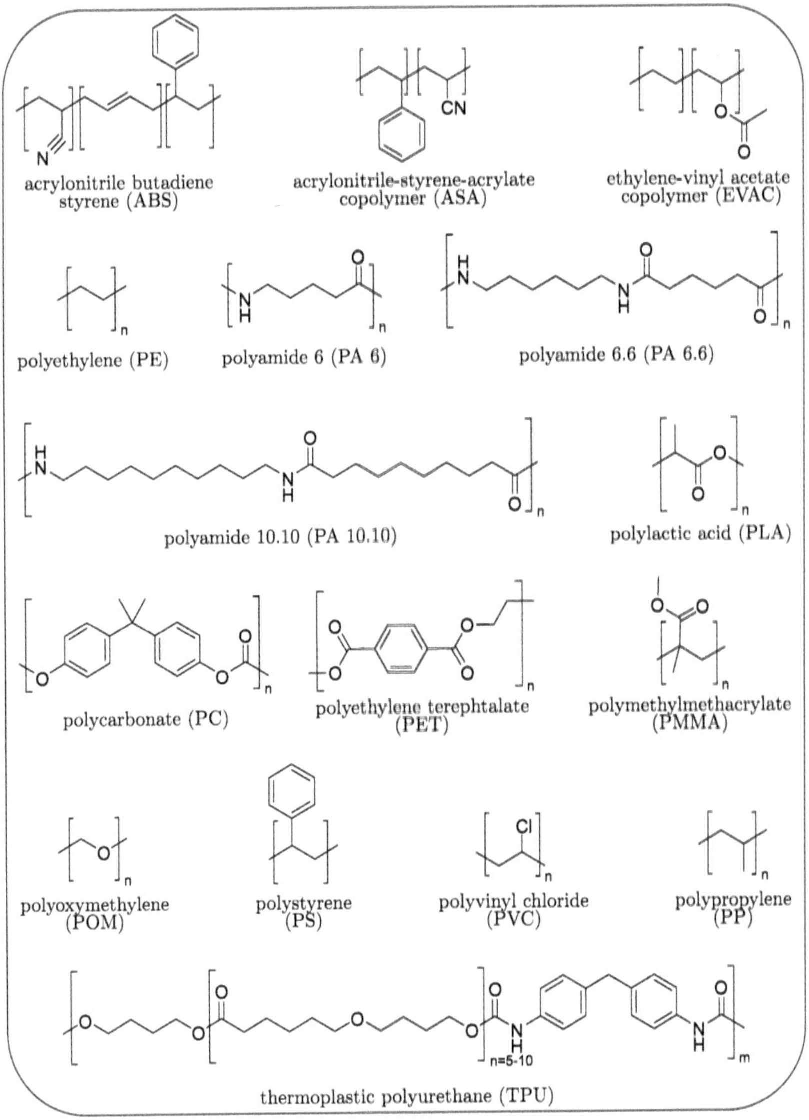

acrylonitrile butadiene styrene (ABS)

acrylonitrile-styrene-acrylate copolymer (ASA)

ethylene-vinyl acetate copolymer (EVAC)

polyethylene (PE)

polyamide 6 (PA 6)

polyamide 6.6 (PA 6.6)

polyamide 10.10 (PA 10.10)

polylactic acid (PLA)

polycarbonate (PC)

polyethylene terephtalate (PET)

polymethylmethacrylate (PMMA)

polyoxymethylene (POM)

polystyrene (PS)

polyvinyl chloride (PVC)

polypropylene (PP)

thermoplastic polyurethane (TPU)

Fig. 1.10 Polymers used for the experiments in this thesis

1.1.2.1 Environmental Effects of MP

MPs are ubiquitous in natural ecosystems, including oceans, rivers, soils, and even atmospheric systems.[73–76] Their persistence is primarily attributed to the durable chemical structure of polymers, which resists natural degradation.[79] In soil ecosystems, MPs can alter physical properties such as water retention and soil aggregation, which can hinder plant growth and microbial activity. In aquatic systems, MPs serve as vectors for adsorbed pollutants, including heavy metals, hydrophobic organic contaminants, and pathogens, which can leach into the environment and cause ecotoxic effects.[80–82]

1.1.2.2 Health Effects of MPs on the Human Body

Humans are exposed to MPs primarily through the ingestion of contaminated seafood, bottled water, and salt, as well as through inhalation of airborne particles. [83] Studies have detected MPs in human stool samples and placental tissue, raising concerns about systemic exposure and potential health risks.[84] Ingested MPs can induce inflammation in the gastrointestinal tract, impair nutrient absorption, and disrupt gut epithelial barriers. Furthermore, adsorbed toxic substances such as pesticides, polycyclic aromatic hydrocarbons, and heavy metals may desorb in the gastrointestinal tract, exhibiting to chemical toxicity. Smaller MPs and nanoplastics have the potential to cross the gut barrier, entering the circulatory and lymphatic systems, where they may accumulate in tissues and induce oxidative stress, inflammation, and cytotoxic effects.[85–87] However, significant knowledge gaps remain, necessitating further research.[88]

1.1.2.3 Acrylonitrile Butadiene Styrene (ABS)

Acrylonitrile butadiene styrene (ABS) is a widely used thermoplastic polymer recognized for its excellent mechanical strength, impact resistance, and versatility. With a global production capacity of several million tonnes annually, ABS finds extensive use across multiple industries due to its properties and broad applicability. In the automotive sector, ABS is frequently used to produce dashboards, wheel covers, and interior components, prized for its durability and aesthetic quality. In consumer electronics, it serves as the housing material for televisions, computers, and mobile phones, offering both strength and easy moldability. Household appliances such as vacuum cleaners and kitchen tools benefit from ABS's toughness and resistance to heat. A well-known example is its use in LEGO™ bricks since 1963, thanks to its rigidity and color retention. It is also popular in 3D printing, particularly in fused deposition modeling (FDM), because of its thermal stability and strength. A schematic repeat unit of ABS is given in Fig. 1.11.

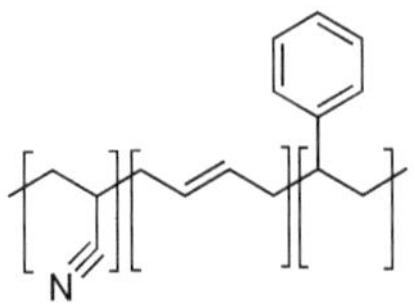

Fig. 1.11 Structure of acrylonitrile butadiene styrene (ABS)

ABS delivers high tensile strength, excellent impact resistance (even at low temperatures), and reliable thermal stability, with a glass transition temperature near 105 °C. It resists aqueous acids, alkalis, and oils, though it is more susceptible to damage from esters, ketones, and some hydrocarbons. Its properties as a good electrical insulator further expand its applications in electronic and electrical products. Despite its advantages, ABS poses certain environmental considerations. Although it is recyclable, not all facilities accept it. When burned, it can release carbon monoxide and hydrogen cyanide. Prolonged exposure to ultraviolet radiation may degrade the polymer, causing discoloration and brittleness.[89, 90]

1.1.2.4 Acrylonitrile-Styrene-Acrylate Copolymer (ASA)

Acrylonitrile-styrene-acrylate (ASA) is an impact-resistant, weatherable thermoplastic often used as an alternative to Acrylonitrile-Butadiene-Styrene (ABS). Its unique properties make it suitable for outdoor applications and products exposed to harsh environments. ASA is commonly applied in the automotive industry for exterior parts such as mirror housings, grilles, and trims, where UV resistance and impact strength are essential. In construction, it is used for roofing tiles, siding, window profiles, shutters, and fencing materials. It is also found in consumer goods like garden furniture, outdoor equipment, sports equipment, and tool housings. ASA is popular in 3D printing for weather-resistant prototypes and models and is used for enclosures in outdoor electrical devices. The copolymer architecture of ASA is sketched in Fig. 1.12.

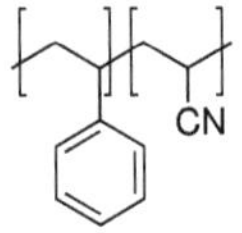

Fig. 1.12 Structure of acrylonitrile-styrene-acrylate (ASA)

ASA has high impact resistance and excellent toughness, even at low temperatures. It is resistant to chemicals such as acids, alkalis, and oils and remains stable against environmental stress cracking. Its thermal properties include high heat resistance with a glass transition temperature of approximately 100 °C. The material demonstrates outstanding UV and weather resistance due to its acrylic component, ensuring colorfastness and gloss retention in outdoor environments. It is easy to process via injection molding, extrusion, and thermoforming and can be painted or dyed without primers. The advantages of ASA include its ability to retain mechanical and aesthetic properties after prolonged exposure to sunlight, moisture, and temperature changes. It offers better chemical stability compared to ABS, does not yellow or degrade in outdoor use, and is easy to fabricate, similar to ABS. However, ASA is more expensive than ABS and has slightly lower impact resistance at high strain rates. ASA was developed to address the limitations of ABS in outdoor environments. It is sometimes blended with other materials to enhance properties like rigidity or processability. The material is recyclable and often used in eco-friendly applications.[91]

1.1.2.5 Ethylene-Vinyl Acetate Copolymer (EVAC)

Ethylene-vinyl acetate copolymer (EVAC) is a versatile thermoplastic polymer composed of ethylene and vinyl acetate. The properties of EVAC depend on the vinyl acetate content, which typically ranges between 10% and 40%. It is widely used due to its flexibility, impact resistance, and adhesive qualities. EVAC is commonly applied in various industries. In packaging, it is used for films, shrink wraps, and sealants, providing excellent transparency and sealing performance. In footwear, it is the primary material for midsoles and outsoles, offering cushioning, flexibility, and durability. It is also utilized in adhesives, particularly in hot-melt adhesives, due to its strong bonding properties. In construction, it serves as a component in waterproofing membranes, roof coatings, and carpet backings. Additionally, EVAC is a popular material for toys and sporting goods because of its softness and resilience. It is also used in photovoltaic panels as an encapsulant to protect solar cells. The ethylene–vinyl acetate motif appears in Fig. 1.13.

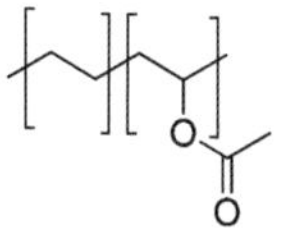

Fig. 1.13 Structure of ethylene-vinyl acetate copolymer (EVAC)

The properties of EVAC include excellent flexibility, elasticity, and toughness, even at low temperatures. It is resistant to water, oils, and certain chemicals, and it has good weatherability and UV resistance. EVAC also has a low melting point, making it easy to process through methods such as extrusion, injection molding, and blow molding. However, its thermal stability is limited, and it may degrade at higher temperatures. The advantages of EVAC include its versatility, cost-effectiveness, and ability to be tailored for specific applications by adjusting the vinyl acetate content. It provides a combination of softness and strength, along with good adhesion to a variety of materials. While its thermal stability is lower than some other polymers, its broad range of uses and excellent performance make it an indispensable material in numerous industries.[92]

1.1.2.6 Polyethylene (PE)

Polyethylene (PE) is one of the most widely used thermoplastic polymers globally, derived from the polymerization of ethylene. It is available in several grades and types, including low-density polyethylene (LDPE), high-density polyethylene (HDPE), linear low-density polyethylene (LLDPE), and ultra-high-molecular-weight polyethylene (UHMWPE). Its properties and applications vary based on density and molecular structure. Polyethylene is extensively used in packaging, such as plastic bags, films, and containers, due to its flexibility, lightweight, and moisture resistance. In construction, it is applied in pipes, insulation, and geomembranes because of its durability and chemical resistance. The automotive industry utilizes polyethylene for fuel tanks, dashboards, and liners, while in consumer goods, it is used for household items, toys, and kitchenware. Additionally, polyethylene is employed in medical applications, such as prosthetics and medical tubing, due to its biocompatibility and toughness. The molecular structure is depicted in Fig. 1.14.

Fig. 1.14 Structure of polyethylene (PE)

The properties of polyethylene include excellent chemical resistance, low moisture absorption, and good electrical insulating characteristics. LDPE is known for its flexibility and impact resistance, while HDPE is valued for its strength, rigidity, and high impact resistance. Polyethylene has a low melting point, making it easy to process via extrusion, injection molding, blow molding, and rotational molding. However, it has limited thermal and UV stability, which can be improved with additives. Polyethylene is advantageous due to its versatility, cost-effectiveness, and recyclability. Its lightweight nature, combined with its resistance to moisture, chemicals, and environmental stress, makes it suitable for a wide range of applications. While it has some limitations, such as susceptibility to UV degradation and relatively low mechanical strength at high temperatures, its adaptability and ease of production have solidified its role as a critical material in modern manufacturing and everyday products.[93]

1.1.2.7 Polyamide 6 (PA 6)

Polyamide 6 (PA 6), also known as Nylon 6, is a synthetic thermoplastic polymer made via ring-opening polymerization of caprolactam. It is known for its excellent mechanical properties, toughness, and chemical resistance, making it one of the most widely used engineering plastics. Polyamide 6 is extensively utilized in the automotive industry for components like gears, bearings, and fuel lines, thanks to its high strength, wear resistance, and ability to withstand elevated temperatures. In textiles, it is a key material for fibers used in clothing, ropes, and nets because of its durability and elasticity. In industrial applications, it is used for conveyor belts, hoses, and bushings, where toughness and chemical resistance are critical. Additionally, it is applied in packaging films, offering excellent barrier properties against gases and odors. The Nylon 6 repeat unit is shown in Fig. 1.15.

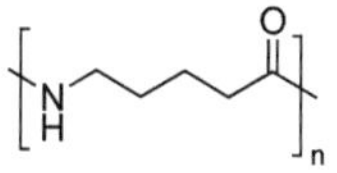

Fig. 1.15 Structure of polyamide 6 (PA 6)

The properties of polyamide 6 include high tensile strength, impact resistance, and good thermal stability. It has excellent resistance to oils, greases, and many chemicals, though it is sensitive to strong acids and bases. PA 6 is hygroscopic, meaning it absorbs moisture from the environment, which can affect its mechanical and dimensional properties. However, this also contributes to its flexibility and toughness. Polyamide 6 is easy to process via injection molding, extrusion, and blow molding. It is commonly modified with fillers like glass fibers to enhance specific properties such as stiffness and dimensional stability. Its melting point is around 220 °C, and it retains good mechanical performance at elevated temperatures.[94]

1.1.2.8 Polyamide 6.6 (PA 6.6)

Polyamide 6.6 (PA 6.6), also known as Nylon 6.6, is a high-performance thermoplastic polymer produced by the condensation polymerization of hexamethylene diamine and adipic acid. It is widely used in the automotive industry for parts such as engine covers, gears, radiator tanks, and air intake manifolds, thanks to its high strength, stiffness, and heat resistance. In textiles, it is used to make durable fibers found in carpets, upholstery, and clothing. Industrial applications include conveyor belts, hoses, and bushings, while in electronics, it is utilized for connectors and housings because of its electrical insulating properties. The molecular structure is shown in Fig. 1.16

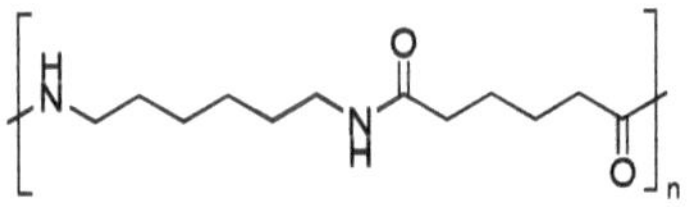

Fig. 1.16 Structure of polyamide 6.6 (PA 6.6)

The material exhibits high tensile strength, excellent wear and abrasion resistance, and good chemical resistance to oils, fuels, and greases. Its melting point of approximately 265 °C is higher than that of polyamide 6, making it suitable for high-temperature applications. Polyamide 6.6 is hygroscopic, absorbing moisture from the environment, which can influence its mechanical properties and dimensional stability. It retains toughness and flexibility even after moisture absorption. Polyamide 6.6 is processed through injection molding, extrusion, and blow molding. It is often reinforced with fillers such as glass fibers or carbon fibers to improve properties like rigidity, dimensional stability, and thermal resistance. It can also be modified with additives for UV protection or flame retardancy, enabling its use in a broader range of applications.[95]

1.1.2.9 Polyamide 10.10 (PA 10.10)

Polyamide 10.10 (PA 10.10) is a bio-based thermoplastic polymer derived from sebacic acid which can be obtained from castor oil and decamethylenediamine. It is known for its excellent chemical resistance, low moisture absorption, and high durability, making it suitable for a range of industrial and consumer applications. PA 10.10 is widely used in the automotive industry for fuel lines, brake hoses, and pneumatic tubes due to its excellent resistance to chemicals and hydrocarbons.

In industrial applications, it is utilized for conveyor belts, tubing, and seals where chemical resistance and mechanical strength are required. In the consumer sector, it is used for sports equipment, toothbrush bristles, and durable textiles. In Fig. 1.17 the molecular structure can be observed.

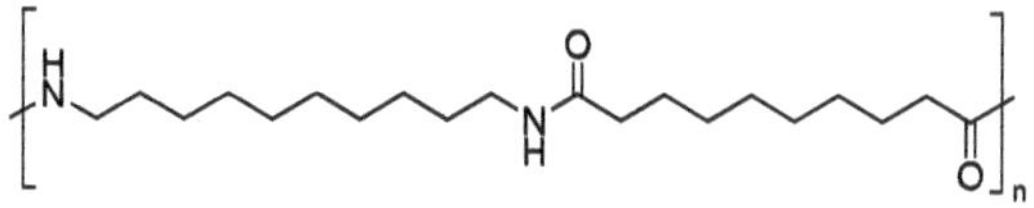

Fig. 1.17 Structure of polyamide 10.10 (PA 10.10)

The properties of polyamide 10.10 include a lower density and better flexibility compared to many other polyamides, along with high chemical resistance to oils, greases, fuels, and various solvents. Its moisture absorption is significantly lower than traditional polyamides like PA 6 or PA 6.6, which helps maintain dimensional stability and mechanical performance in humid environments. The melting point of PA 10.10 is around 215 °C, and it has good thermal stability for medium-temperature applications. Polyamide 10.10 is processed through methods like injection molding, extrusion, and blow molding. It is often used in reinforced grades with glass fibers or other fillers to enhance specific properties such as stiffness and thermal performance. The bio-based origin of sebacic acid makes PA 10.10 a more sustainable alternative compared to fully petroleum-derived polyamides. Its combination of chemical resistance, low moisture uptake, and mechanical durability ensures its utility in a variety of demanding environments.[96, 97]

1.1.2.10 Polycarbonate (PC)

Polycarbonate (PC) is a high-performance thermoplastic polymer made from bisphenol A (BPA) and phosgene. It is known for its exceptional toughness, transparency, and impact resistance, making it a versatile material for a wide range of applications. Polycarbonate is used in the construction industry for glazing and roofing panels, offering high durability and transparency as an alternative to glass. In the automotive industry, it is applied in headlight lenses, interior components, and glazing systems. In consumer electronics, polycarbonate is used for smartphone cases, laptop housings, and compact disc (CD/DVD) production. It is also extensively used in safety equipment such as helmets, goggles, and bulletproof windows. Medical applications include devices, surgical instruments, and protective barriers. The bisphenol A polycarbonate linkage is depicted in Fig. 1.18.

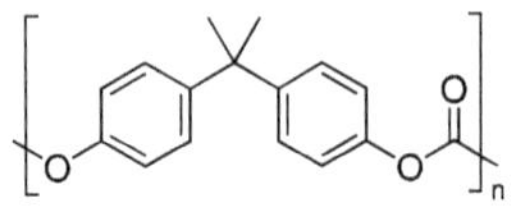

Fig. 1.18 Structure of polycarbonate (PC)

The properties of polycarbonate include high impact resistance, excellent optical clarity, and good thermal stability. It has a heat deflection temperature of around 130 °C and remains dimensionally stable over a broad temperature range. Polycarbonate is resistant to oils, greases, and weak acids but is sensitive to strong bases, solvents, and UV radiation, which can cause degradation. UV stabilizers can be added to enhance its weatherability. Polycarbonate is easily processed through injection molding, extrusion, and thermoforming. It can also be machined and laser-cut for precision applications. The material is commonly blended with other polymers or reinforced with fillers to enhance specific properties, such as rigidity, chemical resistance, or flame retardancy.[98]

1.1.2.11 Polyethylene Terephthalate (PET)

Polyethylene terephthalate (PET) is a thermoplastic polymer from the polyester family, synthesized through the polycondensation of ethylene glycol and terephthalic acid. PET is widely used due to its strength, chemical resistance, and excellent barrier properties. It is primarily used in packaging, particularly for bottles and containers for beverages, food, and personal care products. It is also employed in textile applications under the name polyester, where it is used to produce fabrics, ropes, and tire reinforcements. In industrial applications, PET is utilized for films, thermoformed trays, and components such as electrical insulators. Its recyclability also makes it popular for sustainable packaging solutions and fiber production from recycled materials. The terephthalate ester repeat unit is shown in Fig. 1.19.

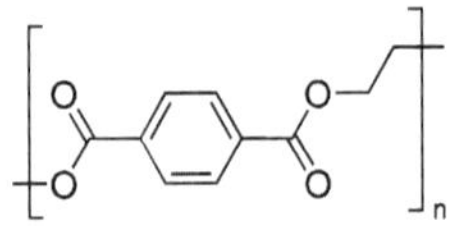

Fig. 1.19 Structure of polyethylene terephthalate (PET)

The properties of PET include high tensile strength, impact resistance, and excellent clarity. It has good chemical resistance to acids, oils, and alcohols, as well as low permeability to gases and moisture, making it ideal for food and beverage containers. PET exhibits good thermal stability, with a melting point around 260 °C, and is easily processed by injection molding, extrusion, blow molding, and thermoforming. However, it is sensitive to strong bases and hydrolytic degradation at elevated temperatures. PET is often modified to enhance specific properties. For example, crystalline PET (C-PET) offers improved thermal resistance, while amorphous PET (A-PET) provides superior clarity.[99]

1.1.2.12 Polylactic Acid (PLA)

Polylactic acid (PLA) is a biodegradable thermoplastic polymer derived from renewable resources such as corn starch, sugarcane, or cassava. It is synthesized through the polymerization of lactic acid or via ring-opening polymerization of lactide. PLA is known for its eco-friendly characteristics and is widely used as an alternative to petroleum-based plastics. PLA is commonly used in the packaging industry for products such as food containers, disposable cups, plates, and cutlery due to its compostability and safety for food contact. In 3D printing, PLA is a popular filament material because of its ease of use, low melting temperature, and minimal warping. The medical industry utilizes PLA for absorbable sutures, drug delivery systems, and implantable devices, since it degrades into lactic acid, a naturally occurring substance in the body. PLA is also used in textiles and films for agricultural and industrial applications. Its polylactide repeat unit is presented in Fig. 1.20.

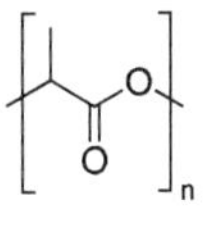

Fig. 1.20 Structure of polylactic acid (PLA)

The properties of PLA include good mechanical strength, high stiffness, and excellent transparency. It has a melting point between 130 °C and 180 °C, depending on its crystalline structure, and a glass transition temperature around 60 °C. PLA is resistant to oils and greases but has limited resistance to high temperatures, UV radiation, and moisture, which can affect its stability and mechanical properties. PLA is processed through injection molding, extrusion, blow molding, and thermoforming. Its biodegradability depends on the conditions, requiring industrial composting facilities with high temperatures and humidity for efficient degradation.

PLA can also be blended with other bioplastics or additives to improve its impact strength, thermal resistance, and durability.[100]

1.1.2.13 Polymethylmethacrylate (PMMA)

Polymethylmethacrylate (PMMA), also known as acrylic or Plexiglas, is a transparent thermoplastic polymer made by the polymerization of methyl methacrylate (MMA). PMMA is known for its excellent optical clarity, weather resistance, and lightweight nature, making it a versatile alternative to glass. PMMA is widely used in the construction industry for windows, skylights, and partitions due to its high transparency and impact resistance. In the automotive industry, it is applied in light covers, instrument panels, and glazing. It is commonly used in signage, display stands, and advertising materials because of its ease of fabrication and ability to be colored. PMMA also finds applications in medical devices such as intraocular lenses, bone cement, and dental prosthetics, as well as in consumer goods like aquariums, furniture, and optical lenses. The methyl methacrylate repeat unit appears in Fig. 1.21.

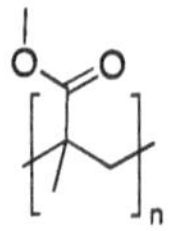

Fig. 1.21 Structure of polymethylmethacrylate (PMMA)

The properties of PMMA include excellent optical clarity with a light transmittance of around 92%, good scratch resistance, and high weatherability. It is resistant to UV radiation, ensuring long-term performance in outdoor applications. PMMA has moderate impact resistance, higher than glass but lower than polycarbonate. Its melting point is approximately 160 °C, and it has a glass transition temperature around 105 °C. PMMA is resistant to many chemicals but can be attacked by strong solvents and alkalis. It is processed through methods like injection molding, extrusion, and thermoforming. PMMA can be machined, laser-cut, or polished to achieve desired shapes and finishes, and additives may be introduced to enhance properties such as impact resistance, heat resistance, or UV stability.[101]

1.1.2.14 Polyoxymethylene (POM)

Polyoxymethylene (POM), also known as acetal or polyacetal, is a high-performance engineering thermoplastic polymer synthesized through the polymerization of formaldehyde. It is known for its excellent mechanical properties, low friction, and high dimensional stability, making it ideal for precision parts in demanding applications. POM is widely used in the automotive industry for components such as fuel system parts, gears, bearings, and door handles due to its strength, stiffness, and wear resistance. In industrial applications, it is used for conveyor belts, sprockets, and pump components because of its low coefficient of friction and excellent sliding properties. In consumer goods, POM is applied in zippers, buttons, and mechanical housings. It is also used in electrical applications for connectors, switches, and insulators. The acetal repeat unit of POM is drawn in Fig. 1.22.

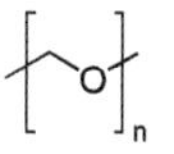

Fig. 1.22 Structure of polyoxymethylene (POM)

The properties of POM include high tensile strength, stiffness, and impact resistance over a wide temperature range. It has excellent resistance to solvents, fuels, and oils, as well as good fatigue resistance and creep performance. POM is self-lubricating, enhancing its wear and abrasion resistance for moving parts. Its melting point is approximately 175 °C, and it is stable in temperatures up to 115–120 °C for continuous use. However, POM is sensitive to strong acids and bases and has limited UV resistance, which can be mitigated with additives. POM is processed through injection molding, extrusion, and machining. It is available in two main forms: homopolymer (POM-H) and copolymer (POM-C). POM-H has slightly higher strength and hardness, while POM-C offers better thermal stability and chemical resistance.[102]

1.1.2.15 Polystyrene (PS)

Polystyrene (PS) is a synthetic thermoplastic polymer made by the polymerization of styrene monomers. It is widely used due to its versatility, cost-effectiveness, and ease of processing. Polystyrene is available in several forms, including solid polystyrene, expanded polystyrene (EPS), and high-impact polystyrene (HIPS), each suited for different applications. Solid polystyrene is commonly used in packaging, disposable cutlery, CD and DVD cases, and laboratory equipment due to its rigidity and transparency. Expanded polystyrene (EPS), a foam version, is widely used for insulation in construction, protective packaging, and disposable food containers. High-impact polystyrene (HIPS), which is blended with rubber, is utilized in applications requiring better impact resistance, such as refrigerator liners, toys, and casings for electronics. The molecular structure is shown in Fig. 1.23.

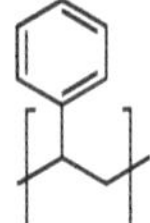

Fig. 1.23 Structure of polystyrene (PS)

The properties of polystyrene include good dimensional stability, excellent clarity (in its general-purpose form), and rigidity. It has a relatively low melting point of approximately 240 °C and is lightweight, making it easy to process through injection molding, extrusion, and thermoforming. However, polystyrene is brittle in its pure form, sensitive to UV light, and has limited resistance to chemicals like strong solvents.[73]

1.1.2.16 Polyvinyl Chloride (PVC)

Polyvinyl chloride (PVC) is a versatile thermoplastic polymer made by the polymerization of vinyl chloride monomers. It is one of the most widely produced plastics globally, known for its durability, chemical resistance, and affordability. PVC is available in two main forms: rigid (RPVC) and flexible, each tailored for specific applications. Rigid PVC (RPVC) is used in construction for pipes, fittings, window frames, and siding due to its strength, weather resistance, and low cost. Flexible PVC is achieved by adding plasticizers, making it suitable for applications such as cables, hoses, flooring, and medical devices like intravenous bags and tubing. PVC is also commonly used in signage, automotive components, and packaging materials, including bottles and shrink wraps. The vinyl chloride repeat unit is depicted in Fig. 1.24.

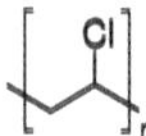

Fig. 1.24 Structure of polyvinyl chloride (PVC)

PVC exhibits excellent chemical resistance to acids, alkalis, and salts. It has good mechanical strength and is flame-resistant, as it contains chlorine, which acts as a natural fire retardant. The material is thermally stable up to 60–80 °C in continuous use, though additives can enhance its thermal resistance. PVC is UV-sensitive but can be stabilized with UV absorbers for outdoor applications. PVC is processed through various methods, including extrusion, injection molding, blow molding, and calendaring. Additives such as stabilizers, plasticizers, and impact modifiers are often incorporated to enhance its properties, such as flexibility, impact resistance, or thermal stability.[103]

1.1.2.17 Polypropylene (PP)

Polypropylene (PP) is a thermoplastic polymer produced through the polymerization of propylene monomers. It is one of the most widely used plastics globally, known for its versatility, durability, and cost-effectiveness. Polypropylene is extensively used in packaging, including food containers, bottle caps, and films, due to its lightweight, moisture resistance, and food safety. In the automotive industry, it is utilized for parts like bumpers, dashboards, and battery cases because of its toughness and resistance to chemicals. It is also used in textiles for ropes, carpets, and nonwoven fabrics such as surgical masks and diapers. In the industrial sector, polypropylene is applied to pipes, storage tanks, and chemical-resistant coatings. The propylene repeat unit is shown in Fig. 1.25.

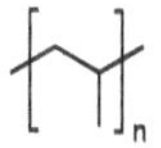

Fig. 1.25 Structure of polypropylene (PP)

The properties of polypropylene include high chemical resistance to acids, bases, and organic solvents, along with excellent tensile strength and impact resistance. It is lightweight, thermally stable up to around 100–130 °C, and has a low moisture absorption rate. Polypropylene is semi-crystalline, giving it a good balance of flexibility and rigidity. However, it is sensitive to UV light and oxidation, which can degrade its properties over time, although stabilizers and additives can mitigate this. Polypropylene is processed through methods such as injection molding, extrusion, blow molding, and thermoforming. It can be modified with fillers, reinforcements, or copolymerization with other monomers to enhance specific properties, such as impact resistance or transparency.[104]

1.1.2.18 Thermoplastic Polyurethane (TPU)

Thermoplastic polyurethane (TPU) is a versatile and durable elastomer belonging to the thermoplastic family of polymers. TPU is created by the reaction of a diisocyanate, a polyol, and a chain extender. It combines the elasticity of rubber with the strength and processability of thermoplastics, making it highly adaptable for a wide range of applications. TPU is widely used in the footwear industry for soles and midsoles due to its flexibility, abrasion resistance, and cushioning properties. In consumer electronics, it is utilized for protective cases, cable coatings, and flexible connectors. The automotive industry employs TPU for interior and exterior components, hoses, seals, and gaskets. Industrial applications include conveyor belts, hoses, and flexible tubing, while medical uses involve tubing, catheters, and wearable devices. TPU is also popular in 3D printing for flexible filaments. A representative urethane linkage for TPU is illustrated in Fig. 1.26.

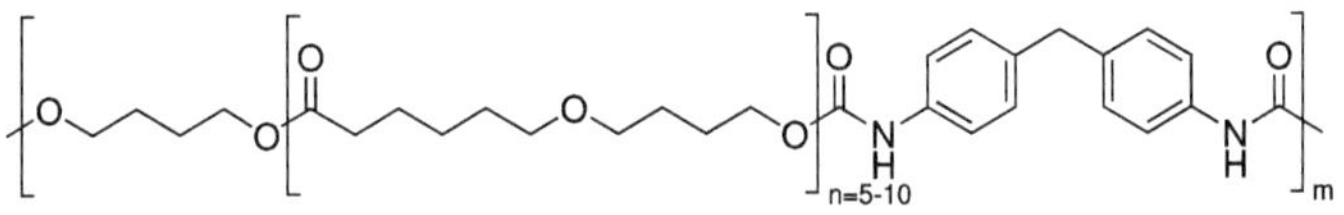

Fig. 1.26 Structure of thermoplastic polyurethane elastomers (TPU)

TPU offers excellent mechanical properties, including high tensile strength, tear resistance, and abrasion resistance. It is highly elastic, retaining flexibility at low temperatures and showing good resilience under repeated stress. TPU is resistant to oils, greases, and many chemicals, as well as weather and UV exposure. Its hardness can be adjusted by altering the ratio of hard and soft segments in its structure, enabling a wide range of flexibility and rigidity. TPU has a melting point around 200 °C and a glass transition temperature below −50 °C. It is processed through methods like injection molding, extrusion, blow molding, and 3D printing. Its inherent versatility allows it to be blended with other polymers or modified with additives to enhance properties such as UV resistance, flame retardancy, or color stability.[105]

1.1.3   MPs as Environmental Pollutants

Microplastics (MPs) have emerged as pervasive environmental pollutants, characterized by their persistence and widespread presence in various ecosystems. These small plastic particles, which originate from the breakdown of larger plastic items or are directly introduced into the environment, are now detected in soils, freshwater systems, and even deep ocean sediments.[75, 76, 79, 106–109] What makes MPs particularly concerning is their ability to act as carriers for other pollutants. Their sometimes hydrophobic surfaces readily adsorb organic and inorganic contaminants, including PFAS. This interaction amplifies the mobility and bioavailability of harmful substances, as MPs can transport these adsorbed pollutants across ecosystems and into the food chain. The combination of PFAS and MPs represents a compounded threat, which warrants a deeper exploration of their interactions.[81, 83, 110, 111]

1.1.4 MPs and PFAS

The interactions between MPs and PFAS involve complex mechanisms driven by environmental conditions, such as pH, temperature, and salinity, as well as the intrinsic properties of both pollutants.[111–113] Advanced analytical methods, including LC-MS/MS, have elucidated these adsorption processes, offering insights into the role of MPs as vectors for PFAS. Understanding these interactions is critical for evaluating their environmental and health impacts and for informing mitigation strategies.[110–112, 114–117]

Field experiments demonstrate significantly higher PFAS adsorption on MPs exposed to natural environments compared to laboratory conditions. This suggests that the presence of inorganic and organic matter in natural ecosystems enhances PFAS retention on MP surfaces.[112] Aging processes, such as UV irradiation and oxidation, further increase the hydrophilic functional groups on MPs, affecting their PFAS adsorption behavior. Environmental factors like salinity and dissolved organic matter can modulate these interactions.[118]

The type of plastic polymer also plays a critical role. For example, virgin PVC exhibited up to 50% adsorption efficiency for PFOS, with adsorption driven by hydrophobic and electrostatic interactions.[111] These interactions facilitate the transport and bioavailability of PFAS, increasing their ecological risks. Moreover, PFAS adsorbed on MPs can desorb in different environmental conditions, such as exposure to sunlight or shifts in pH, further amplifying their mobility.[113]

The combined presence of PFAS and MPs poses significant risks to aquatic organisms and ecosystems. These pollutants can interact synergistically, leading to compounded toxicity effects, such as oxidative stress and immune system dysregulation.[119] Comparative bibliometric studies highlight the growing recognition of these intertwined threats, yet integrated research on their combined impacts remains underexplored.[120]

Their co-occurrence in ecosystems amplifies mobility, bioavailability, and toxicity, necessitating further integrated studies to fully understand and mitigate their impacts on ecosystems and human health. Understanding these interactions is not only crucial for predicting their combined effects but also for developing targeted strategies to reduce their environmental and health consequences effectively.

1.1.5 Liquid Chromatography-Tandem Mass Spectrometry (LC-MS/MS)

Liquid chromatography-tandem mass spectrometry (LC-MS/MS) combines the separation capabilities of liquid chromatography (LC) with the detection power of tandem mass spectrometry (MS/MS), achieving highly specific and sensitive analyses.

Initially, LC separates mixtures in a chromatography column based on each components affinity to the column material. Following separation, techniques such as electrospray ionization (ESI) or atmospheric pressure chemical Ionization (APCI) ionize the components. These ionized molecules are then analyzed by the first mass spectrometer (MS), which selects specific ions based on their mass-to-charge ratio (m/z). After this, selected ions pass into a collision cell where they are fragmented further. The resulting smaller ion fragments are then analyzed by the second MS.[121, 122]

LC-MS/MS offers numerous advantages: it provides high sensitivity and specificity for detecting trace substances within complex mixtures and allows precise quantitative analysis of a wide range of compounds. These attributes make LC-MS/MS an invaluable tool in fields such as pharmacology, environmental science, and biotechnology. The technology also delivers rapid results and detailed information about the analytes. However, drawbacks include the cost of the equipment and the complex, time-consuming sample preparation required.[123]

The operation of an LC-MS/MS system in ESI negative mode is illustrated in Fig. 1.27. In this configuration, samples containing the analyte (M, mass) are introduced into an ultrahigh performance liquid chromatography (UHPLC) column. After separation, the analyte is ionized using electrospray ionization (ESI) in negative mode, which typically involves the loss of a hydrogen ion (H^+), producing negatively charged ions. These ions travel through the extractor cone to the MS/MS analyzer. The system uses two quadrupole filters/mass analyzers (Q1 and Q3) surrounding a collision cell (Q2) filled with argon gas. This setup allows the precursor ion ($[M - H]^-$) to be fragmented into multiple product ions with only selected product ions directed towards the detector for enhanced specificity and accuracy of the analysis.[121, 122]

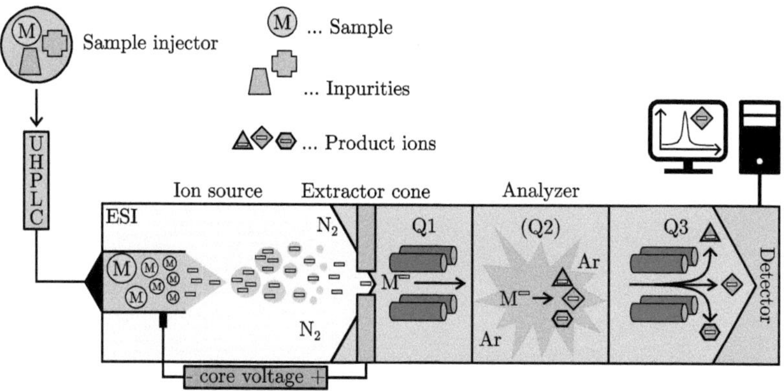

Fig. 1.27 Schematic illustration of a LC-MS/MS instrument in negative ESI mode

1.1.5.1 Multiple Reaction Monitoring (MRM)

MRM operates by selecting specific precursor ions and fragmenting them into product ions. In the tandem mass spectrometer (MS/MS) the following quadrupoles are present:

First Quadrupole (Q1):
Acts as a mass filter which selects precursor ions of a specific mass-to-charge ratio m/z that are associated with the analyte of interest.

Collision Cell (Q2):
Selected precursor ions are collided with a gas, resulting in a fragmentation into product ions. In Q2 the collision energy is precisely regulated to induce fragmentation that specifically aligns with the properties of the analyte.

Second Quadrupole (Q3):
Acts as another mass filter, however, it only selects specific product ions for detection. Only product ions that match the expected m/z values pass through to the detector.

Thanks to its numerous benefits, including high sensitivity, specificity, accurate quantification, and multiplexing capabilities, this mode is extensively employed in various applications.[121, 122]

Experimental

2

2.1 Materials and Methods

This thesis is based on preliminary findings and methodologies developed in collaboration with Johanna Freilinger, Christoph Kappacher, Konstantin Huter, Jan Back, Thomas S. Hofer, Christian W. Huck and Rania Bakry with the provisional title: *Interactions between Perfluorinated Alkyl Substances (PFAS) and Microplastics (MPs): Findings from an Extensive Investigation.* The manuscript of the published study is included in the appendix and available in the ESM. The experimental approaches and analytical techniques detailed herein are adapted from this collaborative effort.

2.1.1 Chemicals

In Table 2.1 an overview of the utilized chemicals in this work can be observed. Purified water utilized in analyzes was produced using a Merck Millipore Milli-Q™ Reference Ultrapure Water Purification System, with deionized water as the input.

Supplementary Information   The online version contains supplementary material available at https://doi.org/10.1007/978-3-658-49859-7_2.

K. Huter, *Unraveling PFAS-Microplastic Interactions*, BestMasters,
https://doi.org/10.1007/978-3-658-49859-7_2

Table 2.1 List of utilized chemicals with their purity and supplier (including location)

Chemical	Purity	Supplier (Location)
Ammonium acetate	$\geq$ 97%	Carl Roth GmbH + Co. KG (Karlsruhe, Germany)
Ammonium perfluoro(2-methyl-3-oxahexanoate)(Gen-X)	95%	Manchester Organics Ltd (Runcorn, UK)
Ammonium hydroxide solution	25%	Fluka, Sigma-Aldrich GmbH (Steinheim, Germany)
Heptadecafluorooctanesulfonic acid potassium salt (PFOS)	$\geq$ 92%	Sigma-Aldrich GmbH (Steinheim, Germany)
Nonafluorobutane-1-sulfonic acid (PFBS)	97%	Sigma-Aldrich GmbH (Steinheim, Germany)
Perfluorobutanoic acid (PFBA)	Analytical standard	Sigma-Aldrich GmbH (Steinheim, Germany)
Perfluorooctanoic acid (PFOA)	96%	Sigma-Aldrich GmbH (Steinheim, Germany)
Perfluorononanoic acid (PFNA)	97%	Sigma-Aldrich GmbH (Steinheim, Germany)
Perfluoro-pentanoic acid (PFPeA)	Analytical standard	Sigma-Aldrich GmbH (Steinheim, Germany)
Undecafluorohexanoic acid (PFHxA)	Analytical standard	Sigma-Aldrich GmbH (Steinheim, Germany)
Perfluoro-n-($^{13}C_8$) octanoic acid (^{13}PFOA)	> 99%	Wellington Laboratories (Guelph, Ontario, Canada)
Methanol for LC-MS/MS	$\geq$ 99.99%	VWR (Avantor) (Radnor, PA, USA)
Formic acid (FA)	98–100%	Merck KGaA (Darmstadt, Germany)

2.1.2 HPLC Vials

Vials out of polypropylene (PP) were obtained from Agilent Technologies, Santa Clara, CA, USA as well as from Waters Corp, Milford, MA, USA. Amber glass vials were acquired from Labsolute® (Th. Geyer, Renningen, Germany).

2.1.3   Syringe Filters

Syringe filters were acquired, each featuring a different membrane material: cellulose acetate (CA), nylon, polyvinylidene fluoride (PVDF), and polytetrafluoroethylene (PTFE). All filters have a membrane diameter of 13 mm and a pore size of 0.22 μm

2.1.4   Polymers

Polymer samples including acrylonitrile butadiene styrene (ABS), acrylonitrile-styrene-acrylate copolymer (ASA), ethylene-vinyl acetate copolymer (EVAC), high-density polyethylene (HDPE), low-density polyethylene (LDPE), linear low-density polyethylene (LLDPE), polyamide 6 (PA 6), polyamide 6.6 (PA 6.6), polyamide 10.10 (PA 10.10), polycarbonate (PC), polyethylene terephthalate (PET), polylactic acid (PLA), polymethylmethacrylate (PMMA), polyoxymethylene (POM), polystyrene (PS), polypropylene (PP), polyvinyl chloride (PVC), and thermoplastic polyurethane elastomers (TPU) were sourced in pellet form from various manufacturers. Each polymer type was milled individually in a centrifugal mill ZM 200 (Retsch, Verder Scientific, Haan, Germany), using liquid nitrogen for cooling to avoid melting during the process. The milling involved two stages: first, using a ring sieve with 1 mm trapezoid holes, followed by a finer sieve with 0.5 mm holes, resulting in particle sizes approximately between 400 and 500 μm.

2.1.5   LC-MS/MS

The LC-MS/MS analysis was conducted using a Waters Acquity Premier LC system connected to a Waters Xevo TQD triple quadrupole. The setup included a primary separation column, Waters ACQUITY UPLC BEH C18, measuring 2.1x100 mm with 1.7 μm particle size, and a VanGuard™ pre-column. Additionally, a Waters AtlantisTM Premier BEH C18 AX 5 column, 2.1x50 mm with 5 μm particle size, was used as a pre-column to filter out any potential PFAS from the LC system. Column temperatures were maintained at 40 °C and autosampler temperatures at 4 °C, with an injection volume of 10 μL. The gradient solvent system used was 2 mM ammonium acetate in water and a 95/5 (v/v) mixture of MeOH/2 mM ammonium acetate in water, flowing at 0.30 mL min^{-1}. The gradient schedule was set as follows: from 0–1 min (5–25% B), 1–6 min (25–50% B), 6–13 min (50–85% B), 13–14 min (85–95% B), 14–17 min (maintained at 95% B), 17–18 min (95–5% B), finishing with 18–22 min (reverting to 5% B). Data acquisition and tuning were managed with Waters MassLynx and IntelliStart software, respectively. The multi reaction monitoring (MRM) method was executed in negative electro-spray ionization (ESI) mode, employing nitrogen as the sheath and auxiliary gas and argon as the collision gas. Calibration was performed using standards such as PFBA, Gen-X, PFOA, and PFOS. Details on the precursor ions, their transitions, and specific settings like cone voltage and collision energies are documented in Table 2.2. To prevent mass spectrometer contamination from polar compounds like salts, the system column hold-up time was calibrated using a 2 $\mathrm{mg\,L^{-1}}$ thiourea solution in water. MS data acquisition commenced at 1.4 min and concluded at 16 min, with peak integration carried out using TargetLynx software included in MassLynx. The associated chromatogram is depicted in Fig. 2.1.

Table 2.2 Each PFAS as the analyte with associated retention times (R_t) and MS/MS parameters in MRM mode

Analyte	R_t / min	Precursor Ion $[M-H]^-$	Fragment Ion $[M-H]^-$	Cone Voltage / V	Collision Energy / V
PFBA	3.72	213	169	18	14
PFPeA	6.31	263	218, 219	17	7, 9
PFBS	6.91	299	80, 83, 99, 119	56	32, 28, 32, 24
PFHxA	8.40	313	119	18	22
Gen-X	8.88	284	119, 169, 185	38	28, 10, 18
PFOA	11.14	413	169, 369	18	16, 8
^{13}PFOA	11.15	415	170, 219, 370	18	16, 14, 20
PFOS	11.80	499	80, 99, 130	60	46, 50, 48
PFNA	12.09	463	169, 219, 269	20	18, 14, 14

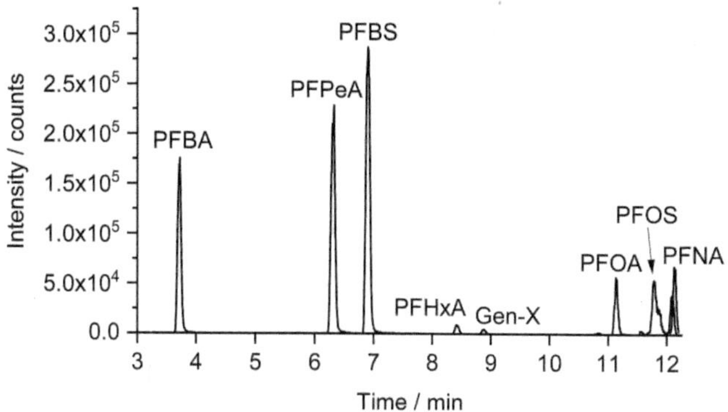

Fig. 2.1 Overview of the eight PFAS in a chromatogram

2.1.6 Calibration

The calibration process was conducted using 250 μL of an internal standard (perfluoro-n-($^{13}C_8$) octanoic acid) with a concentration of $20\,\mu g\,L^{-1}$. Five vials were prepared for the experiment, with their concentrations outlined in Table 2.3. A PFAS mixture, containing the eight compounds listed in subsection 2.1.1 at a concentration of $100\,\mu g\,L^{-1}$ per PFAS, was added to the volumes specified in Table 2.3.

Table 2.3 Vial, concentration and the volume of water (H_2O) and the PFAS mix for the calibration

Vial	Concentration / $\mu g\,L^{-1}$	H_2O / μL	PFAS mix / μL
1	20	200	20
2	40	150	40
3	60	100	60
4	80	50	80
5	100	0	100

Subsequently, the prepared vials were analyzed using LC-MS/MS. The data was processed in MassLynx. The resulting calibration curves are shown in Fig. 2.2. The coefficients of determination (R^2) and equations from Fig. 2.2 are observable in Table 2.4.

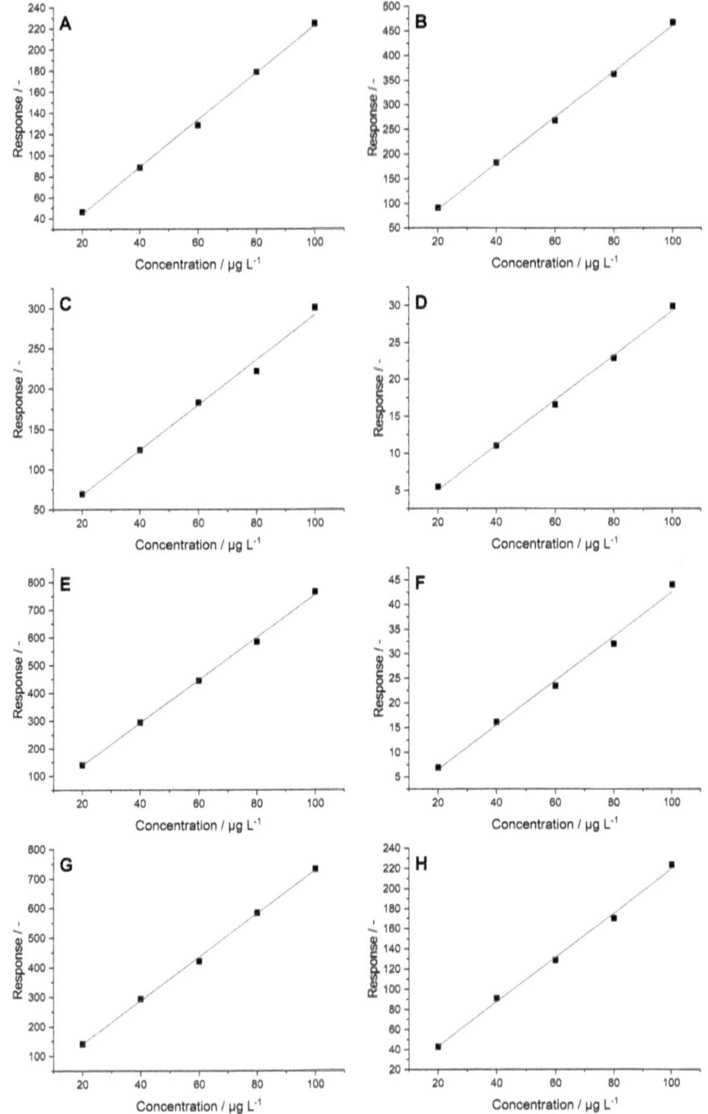

Fig. 2.2 Linear fit of A) PFOA, B) PFBA, C) PFOS, D) Gen-X, E) PFPeA, F) PFBS, G) PFHxA and H) PFNA

Table 2.4 Coefficients of determination (R^2), labels of Fig. 2.2 and linear regression equations for the eight PFAS with the equation $y = a + b \cdot x$, where y is the response, a is the intercept, b is the slope in $L\,\mu g^{-1}$, and x is the concentration in $\mu g\,L^{-1}$

PFAS Compound	Label	R^2	Equation
PFOA	A	0.9984	$y = -0.6611 + 2.23469 \cdot x$
PFBA	B	0.9990	$y = -5.2788 + 4.65684 \cdot x$
PFOS	C	0.9944	$y = 11.6869 + 2.80346 \cdot x$
Gen-X	D	0.9975	$y = -1.0422 + 0.30337 \cdot x$
PFPeA	E	0.9990	$y = -16.5388 + 7.71534 \cdot x$
PFBS	F	0.9949	$y = -2.5094 + 0.45047 \cdot x$
PFHxA	G	0.9990	$y = -7.1921 + 7.36526 \cdot x$
PFNA	H	0.9976	$y = -0.6915 + 2.20084 \cdot x$

2.1.7 Procedure

2.1.7.1 Workflow

As seen in Fig. 2.3 the workflow includes the weighing of exactly 100 mg of MPs into 1.5 mL vials. Then, 1 mL of the PFAS standard stock solution (100 $\mu g\,L^{-1}$) was added to each MP. Afterwards, the suspension was thoroughly mixed using a thermoshaker (24 °C, 1000 rpm) for 24 hours to achieve equilibrium. The PFAS-MP suspension was filtered using a syringe filter (CA, 13 mm, 0.22 μm) for the removal of the polymers and was directly transferred into a clean vial. Of the received solution 250 μL were pipetted into an HPLC vial and thoroughly mixed with 250 μL of internal standard (IS) solution with a concentration of 20 $\mu g\,L^{-1}$. Note, that the measurements via LC-MS/MS were carried out immediately after the described workflow.

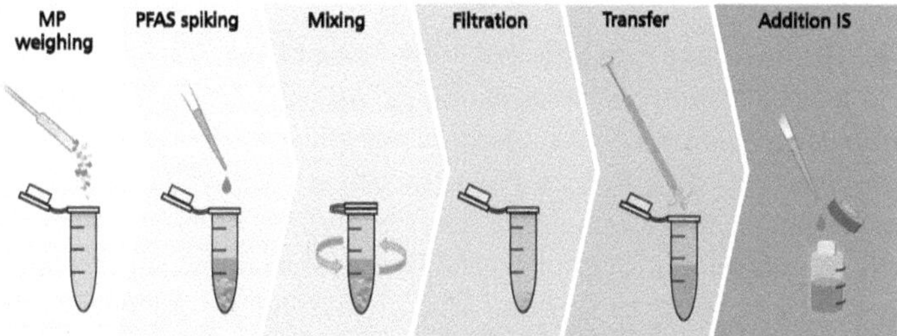

Fig. 2.3 Workflow of the main experiment

2.1.7.2 Calculations

To address unavoidable PFAS losses during the practical workflow, control experiments were conducted in triplicate for each extraction batch. These losses were incorporated into the uptake calculations for each MP polymer to ensure accurate results. The polymer uptake was calculated using Equation (2.1), adjusted to represent the uptake of each PFAS in milligrams per gram of polymer, accounting for aliquotation and dilution factors:

$$Uptake \text{ in } \text{mg}_{\text{PFAS}}\ \text{g}^{-1}\ _{\text{MPs}} = \frac{Conc._{Control} - Conc._{Residue}}{m_{MP}} \tag{2.1}$$

where $Conc._{Control}$ is the average PFAS concentration in the control experiment, $Conc._{Residue}$ is the PFAS concentration after adsorption on the respective MPs, both in mg L^{-1}. The mass of the MP is denoted as m_{MP}, expressed in grams (g). The associated error, combining uncertainties from the control and polymer uptake experiments (both performed in triplicate), was calculated using Equation (2.2). Here, $s_{Control}$ is the standard deviation of the control measurements and $s_{Residue}$ is the standard deviation of the extraction protocol. These values were combined to compute the standard deviation of the uptake (s_{Uptake})

$$s_{Uptake} \text{ in } \text{mg}_{\text{PFAS}}\ \text{g}^{-1}\ _{\text{MPs}} = \sqrt{(s_{Control})^2 + (s_{Residue})^2} \tag{2.2}$$

2.1.7.3 Preliminary Experiment—HPLC Vials

The potential uptake of PFAS by HPLC vials during storage and measurements were investigated. For this, aliquots of 1 mL of a stock solution of each PFAS with a concentration of 100 µg L^{-1} were dispensed into the respective vials. Each experiment was repeated five times. Then the temperature of the vials was controlled by a thermoshaker (24 °C, 1000 rpm) for 24 hours. The detection was carried out as stated in subsection 2.1.5.

2.1.7.4 Preliminary Experiment—Syringe Filter

To assess the potential PFAS losses during the MP-suspensions filtration process, four different syringe filters were used, all identical in diameter and pore size but made from various materials: polytetrafluoroethylene (PTFE), nylon, cellulose acetate (CA), and polyvinylidene fluoride (PVDF). A uniform stock solution was prepared, containing eight targeted PFAS compounds at a concentration of 100 µg L^{-1} each. For the experiment, 1 mL of this solution was drawn into a 1 mL syringe and passed through each filter into a 1 mL Eppendorf vial, this procedure was repeated five times for each filter type. Additionally, to test for any background from the filters, 1 mL of water was passed through each filter, collected in an Eppendorf vial, and analyzed under the same LC-MS/MS conditions. The procedure was repeated for each filter type. Detection of the filtered samples was then conducted using the same parameters as specified in subsection 2.1.5.

2.1.7.5 Long-Term Study

In the experiments, 100 mg of MPs were weighed for each sample, treated with a PFAS mixture, gently shaken, and left undisturbed at room temperature for 30 days. Afterwards, the solution was filtered with CA and transferred into an Eppendorf vial. 250 μL of the long-term solution and of the IS was pipetted into an HPLC vial for LC-MS/MS measurements. For LC-MS/MS analysis, 250 μL of the long-term solution and the IS were pipetted into an HPLC vial. The calibration was performed simultaneously to ensure it was freshly prepared.

2.1.7.6 pH Experiment

The pH 7 solution, consisting solely of water, was prepared as described in sub-subsection 2.1.7.1. The pH meter (WTW pH 540 GLP) was calibrated using technical buffer solutions from WTW. Subsequently, formic acid (FA) and ammonium hydroxide were gradually added to adjust the pH to 4 and 10, respectively. To prevent any pH shifts over time, the LC-MS/MS measurements were carried out immediately after preparation.

2.1.8 Theoretical Calculations

To investigate the interactions between PFAS molecules and MPs, a global optimization (GO) study was conducted using the MACE-OFF23 neural network potential (NNP).[124, 125] Polyethylene (PE) was modeled as a linear $C_{16}H_{34}$ molecule, while polyamide (PA) was represented by one nylon 6.6 fragment capped at the end with one CH_3 group each. PFOA and PFBA were chosen as substrate molecules. Each molecule underwent initial energy minimization using the MACE-OFF23 NNP. The resulting minimum-energy geometry was then centered at the origin, with the main aliphatic chains aligned parallel to the z-axis.

Initial configurations for the global optimization were created by moving the substrate molecule within a ±5 Å range along the x-, y-, and z-axes in 1 Å increments relative to the stationary polymer molecule. To explore different molecular orientations, an additional global optimization was performed for each pair by mirroring the substrate in the xy-plane (equivalent to a 180° rotation relative to the polymer). This resulted in two GO runs per substrate-polymer pair, for a total of eight global optimizations using the MACE-OFF23 NNP. A minimum atom separation of 1.5 Å was maintained to prevent molecular overlap. The number of initial configurations varied between 726 and 796 for each substrate-polymer pair.

After energy minimizations for all initial geometries, the resulting minimum-energy structures were ranked in descending order of total energy. The five lowest-energy conformations per substrate-polymer pair were selected for further minimization using the semi-empirical second-generation extended tight binding (GFN2-xTB) method, implemented in the DFTB+ package.[126–129] This method included solvent effects via the GBSA model (generalized Born model with surface area contributions), with water parameters sourced from the Grimme labs GBSA repository.[130] Interaction energies (E_{Int}) for the lowest-energy conformers were calculated as:

$$E_{Int} = E_{Poly\text{-}Sub} - (E_{Poly} - E_{Sub}) \tag{2.3}$$

where E_{Poly} and E_{Sub} are the minimized energies of the isolated polymer and substrate, and $E_{Poly\text{-}Sub}$ is the energy of the associated interacting system (see Equation (2.3)).

2.1.9 Differential Volume Distribution

Particle size distributions were determined using laser diffraction (Malvern Instruments, Mastersizer 2000). Suspensions of the particles were prepared in ethanol and subjected to ultrasonication to ensure even dispersion. Ethanol was used as the dispersing medium during the measurements. The refractive index was set to 1.36 for the dispersant and ranged from 1.476 to 1.62 for the particles. PMMA was directly dispersed for 60 seconds, while all other particles underwent pre-dispersion by ultrasonication in ethanol for 1 hour.

Results and Discussion

3

3.1 Preliminary Experiments

Preliminary experiments were conducted to address the problematic storage and adsorption characteristics of PFAS.[131] First, three vial brands were evaluated to determine a suitable sample container: polypropylene (PP) vials from Agilent and Waters as well as amber glass vials from Labsolute®. Secondly, the syringe filter materials were tested, each featuring a 13 mm membrane diameter and a 0.22 μm pore size. The subsequent sections detail the findings of these preliminary investigations, which shaped the procedure adopted in the main experiments.

3.1.1 Selecting HPLC Vial Materials for PFAS-Polymer Analysis

Results of the HPLC vial testing (Fig. 3.1) demonstrated clear differences in PFAS recovery depending on the vial material over a 24-hour storage period. All compounds were spiked at a target concentration of $100\,\mu g\,L^{-1}$, indicated by the dotted line. In amber glass vials, the measured concentrations for most PFAS consistently fell below this baseline, suggesting significant adsorption to or interaction with the glass surface and leading to underestimation of PFAS levels. Consequently, glass vials were excluded from further experiments due to analyte losses compromising data accuracy.

By contrast, polypropylene (PP) vial brands achieved recoveries/concentrations at or above 100% for most PFAS compounds. For PFNA, Waters PP vials achieved concentrations above the dotted line ensuring reliable and accurate recovery. On the other hand, Agilent PP vials showed concentrations below the 100% mark for PFNA, indicating less precise quantification. This superior performance of Waters PP vials

© The Author(s), under exclusive license to Springer Fachmedien Wiesbaden GmbH, part of Springer Nature 2025
K. Huter, *Unraveling PFAS-Microplastic Interactions*, BestMasters,
https://doi.org/10.1007/978-3-658-49859-7_3

across PFOS and PFNA providing more accurate and reproducible measurements, further supporting their selection for subsequent experiments. Consequently, Waters PP vials were chosen for quantification, highlighting the critical impact of vial material on analytical accuracy.

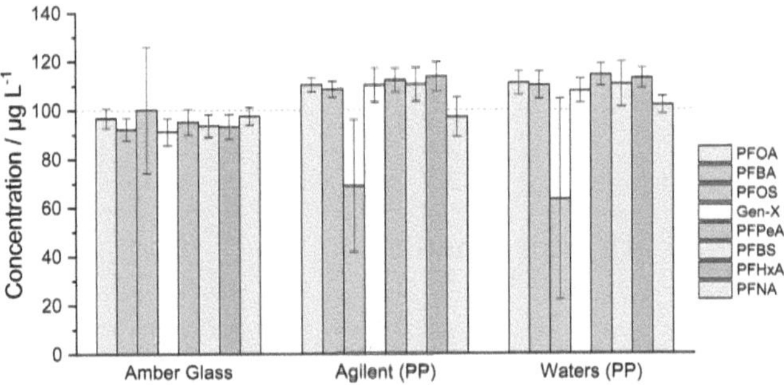

Fig. 3.1 Concentration in $\mu g\,L^{-1}$ of quantified PFAS post-extraction in HPLC Vials over 24 hours. The horizontal dotted reference line marks 100% recovery/concentration

3.1.2   Selecting Syringe Filter Materials for PFAS-Polymer Analysis

The heat map shown in Fig. 3.2 illustrates that the type of filter membrane significantly affects PFAS adsorption. Additionally, short-chain PFAS such as PFBS, PFBA and PFHxA generally exhibit lower uptake than longer-chain PFAS (PFOA, PFNA and PFOS), reflecting their higher polarity and reduced hydrophobic interactions. PFOS is a notable outlier, proving especially difficult to manage under standard laboratory conditions, as already observed in Fig. 3.1. PVDF tends to have moderate uptake for PFOS (around 33%) but is comparatively low for other PFAS. Cellulose acetate (CA) is generally very low across all PFAS (most values are near zero, except PFOS at $\approx$18%), which aligns with the statement that CA was chosen for its minimal PFAS adsorption. Nylon shows very high uptake, especially for PFOS ($\approx$99%), PFNA ($\approx$98%), PFHxA ($\approx$94%), and PFOA ($\approx$95%). PTFE also reaches values up to 100% for PFOS and PFNA but is more moderate for short-chain PFAS. This data aligns with the conclusion that filter material significantly influences PFAS recovery.

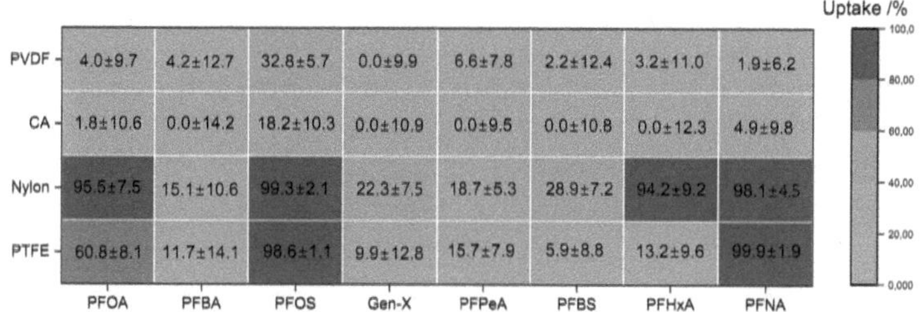

Fig. 3.2 Heat map displaying the adsorption percentages and standard deviation for various combinations of filter materials and PFAS

3.1.3   Differential Volume Distribution of the Polymers

In Fig. 3.3 the differential volume distribution of each polymer can be observed. The measurement results of various polymers is shown in Table 3.1. Most polymers exhibit similar particle sizes, except for PVC-U. This difference likely results from PVC-U being supplied in a pre-finely powdered form, in contrast to the polymers that were milled.

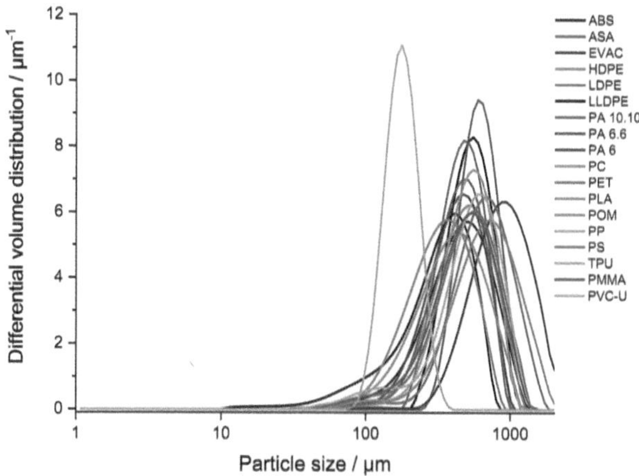

Fig. 3.3 Differential volume distribution of different polymers analyzed by particle size. The x-axis represents particle size (μm), while the y-axis indicates the differential volume distribution (μm^{-1}). Each curve corresponds to a specific polymer, including ABS, ASA, EVAC, HDPE, LDPE, LLDPE, PA 10.10, PA 6.6, PA 6, PC, PET, PLA, POM, PP, PS, TPU, PMMA, and PVC-U

Table 3.1 Differential volume distribution measurements of various polymers obtained using diffraction methods. The table lists the refractive index (RI) and characteristic particle diameters at the 10th ($d(0.1)$), 50th ($d(0.5)$), and 90th ($d(0.9)$) percentiles

Polymer	RI	$d(0.1)$ / μm	$d(0.5)$ / μm	$d(0.9)$ / μm
ABS	1.54	318.08	507.603	775.352
ASA	1.62	278.724	447.737	682.734
EVAC	1.49	194.937	433.419	788.117
HDPE	1.54	230.319	461.924	803.249
LDPE	1.54	222.001	423.611	684.98
LLDPE	1.54	110.46	320.599	553.114
PA 10.10	1.53	389.409	580.845	866.57
PA 6.6	1.53	213.14	412.053	689.669
PA 6	1.53	446.505	828.691	1437.875
PC	1.58	169.937	380.899	722.989
PET	1.52	325.509	622.146	1080.628
PLA	1.48	311.348	650.682	1206.358
POM	1.48	153.566	318.472	569.621
PP	1.49	259.297	486.437	772.089
PS	1.59	222.838	492.614	873.26
TPU	1.55	206.438	521.147	872.08
PMMA	1.48	211.948	471.147	836.657
PVC-U	1.54	119.747	168.262	237.596

3.2   Main Experiments

3.2.1   Time-Dependent Adsorption Behavior

In this study, the adsorption characteristics of four different polymers (PP, LDPE, PA 6.6 and PET) were analyzed over extraction periods of 1, 2, 6, and 24 hours, with a long-term (l.t.) study extending over three weeks, as shown in Fig. 3.4, 3.5, 3.6 and 3.7. The graphs demonstrate that the time required to achieve equilibrium varies by polymer type.

As seen in Fig. 3.4 an initial rise in PFAS uptake at 6 hours, followed by a slight decline at 24 hours and then a renewed increase over the long-term period, is evident for PET. One possible explanation is that the polymer surface initially adsorbs PFAS quickly, leading to higher uptakes at 6 hours. By 24 hours, partial desorption or re-equilibration may occur, particularly if localized saturation on the polymer surface leads to some release back into the surrounding medium, thereby reducing the net measured uptake. Over the extended timeframe, the polymer may experience increased polymer-solvent interaction, allowing PFAS molecules to diffuse more deeply into the polymer's amorphous regions. During this process, solvent or moisture exchange in the matrix can change the polymers porosity or free volume, ultimately enabling a renewed increase in PFAS uptake that becomes apparent in the long-term data. PET possesses ester linkages that add moderate polarity, allowing dipole–dipole or hydrogen-bond interactions, yet it also contains regions that can facilitate hydrophobic interactions, especially with longer-chain PFAS. The result is a slower but persistent accumulation of PFAS over weeks, which is distinct from the rapid equilibrium observed in PA 6.6 and the lower overall sorption in PP and LDPE.

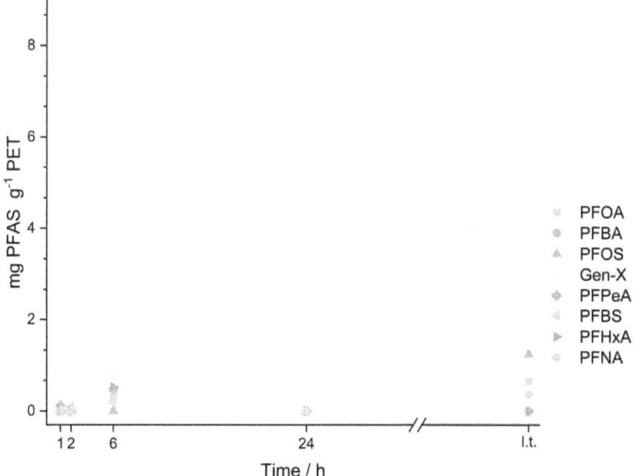

Fig. 3.4 Results of the time-dependent uptake experiment for 1, 2, 6, 24 hours and long-term (l.t.) in mg PFAS per g polymer particles for PET

For polypropylene (PP), the uptake measurements at 1, 2, and 6 hours are shown to be consistent, but a decline at 24 hours can be observed, followed by a slight resurgence in the long-term data as visible in Fig. 3.5. One possible explanation is that PFAS molecules are initially adsorbed onto surface sites, leading to stable readings in the early phase; by 24 hours, localized saturation or minor desorption can temporarily reduce uptake. Over extended periods, slow polymer "wetting" or rearrangements in the polymer matrix may reopen pathways for PFAS, producing a mild increase in the final measurements.

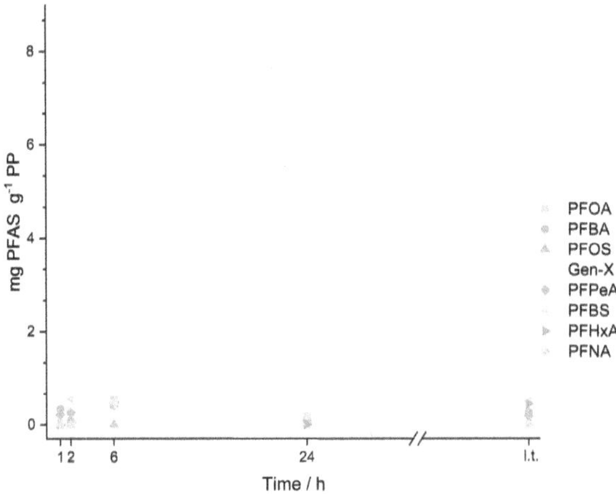

Fig. 3.5 Results of the time-dependent uptake experiment for 1, 2, 6, 24 hours and long-term (l.t.) in mg PFAS per g polymer particles for PP

Low-density polyethylene (LDPE) shows a comparatively high initial PFAS uptake (over 1 mg PFAS per g LDPE) which continues to rise up to 6 hours, then exhibits a slight drop at 24 hours as observed in Fig. 3.6. In the long-term dataset, LDPEs uptake decreases further and ultimately settles at its lowest observed level. This trend could indicate that LDPEs less structured regions readily adsorb PFAS at first, but as the system equilibrates, partial desorption or subtle structural changes, such as reduced porosity or different solvent interactions, cause a net decrease in overall PFAS retention.

PA 6.6, depicted in Fig. 3.7, shows a rapid increase in PFAS uptake, reaching near-complete adsorption within 24 hours for most analytes, with values of around 8 mg PFAS per g of PA 6.6 (equals 100% uptake). Its amide bonds confer relatively high polarity, facilitating hydrogen bonding and electrostatic interactions with both long- and short-chain PFAS. Once these binding sites are occupied subsequent adsorption slows, leading to a stable plateau in uptake that persists through the three-week observation period. This behavior highlights how the polar nature of the polymer backbone can drive strong adsorption.

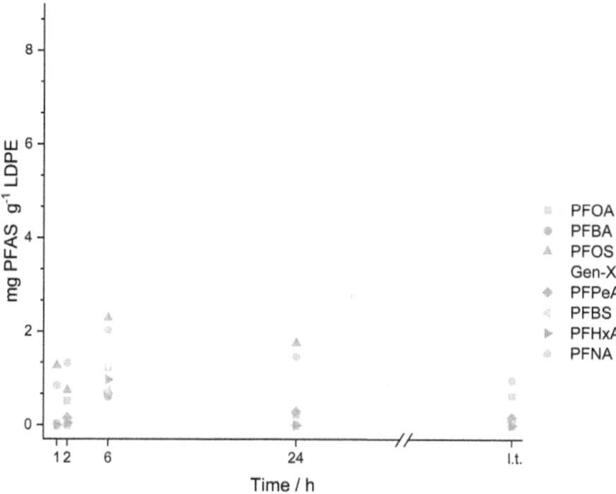

Fig. 3.6 Results of the time-dependent uptake experiment for 1, 2, 6, 24 hours and long-term (l.t.) in mg PFAS per g polymer particles for LDPE

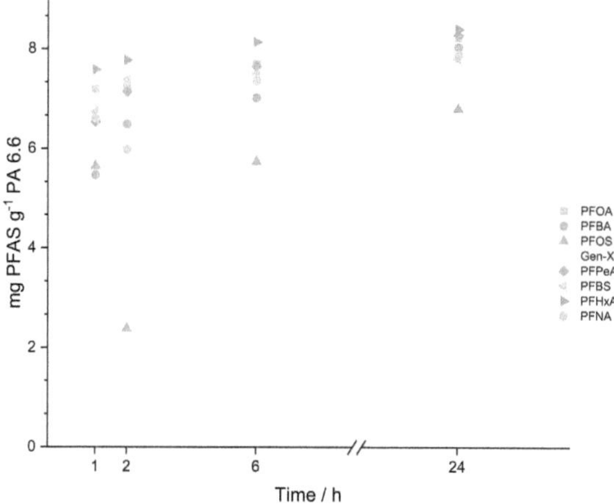

Fig. 3.7 Results of the time-dependent uptake experiment for 1, 2, 6 and 24 hours in mg PFAS per g polymer particles for PA 6.6

To conclude, PET shows initial gains in uptake, followed by partial desorption, and eventually increased uptake over the long-term. PP maintains a mostly steady uptake, with a slight dip observed at 24 hours, before rising again. LDPE exhibits rapid initial uptake, which eventually decreases over time. PA 6.6 achieves near-complete uptake within 24 hours and remains stable afterward. Notably, for the other polymers, including PP, PET, and LDPE, the PFAS uptake initially increases, peaking around the 2-hour mark, before subsequently declining. This behavior may be attributed to the time required for the polymer particles to achieve complete wetting. Despite this initial fluctuation, equilibrium is consistently reached across all tested polymers by 24 hours. This is evident from the stabilization of uptake levels between 6 and 24 hours, remaining within the margin of deviation attributed to measurement errors arising from the extraction process and control measurements. All subsequent experiments will be conducted using a 24-hour time frame for extraction.

3.2.2   Different Polymer Types

The interaction of 18 different polymers with eight PFAS, as illustrated by the radial plots in Fig. 3.8, reveals several noteworthy trends. PAs stand out for their exceptionally high adsorption rates, frequently reaching 100% uptake (equivalent to 8 mg PFAS per gram of polymer). Only minor differences occur among PA 6, PA 6.6, and PA 10.10, although PA 10.10 exhibits notably lower adsorption for PFPeA and PFBA than do the other polyamides. Thermoplastic polyurethane (TPU) and acrylonitrile styrene acrylate (ASA) also show relatively high uptakes, although they do not match the levels seen in the polyamides. By contrast, more widely used polymers such as PET, PP, and PE demonstrate lower adsorption capacities, indicating minimal interactions with PFAS.[132–134]

This marked difference in adsorption behavior cannot be explained by hydrophobicity alone. Instead, it suggests that functional groups, particularly amide linkages, play a important role. Polyamides contain amide bonds that can form hydrogen bonds with the carboxyl or sulfonic acid groups of PFAS. In such interactions, the hydrogen on the polymers amide nitrogen can donate a hydrogen bond to the acid site of PFAS, while the partial positive charge on the amide nitrogen and partial negative charge on the PFAS oxygen further reinforce electrostatic attractions. For carboxylated PFAS, the carbonyl group in the polymer may withdraw electron density from the amide bond, slightly increasing the polarity of the -NH-site, which effectively enhances hydrogen-bond formation. Sulfonated PFAS, featuring strongly acidic sulfonic groups, can similarly engage in hydrogen bonding or

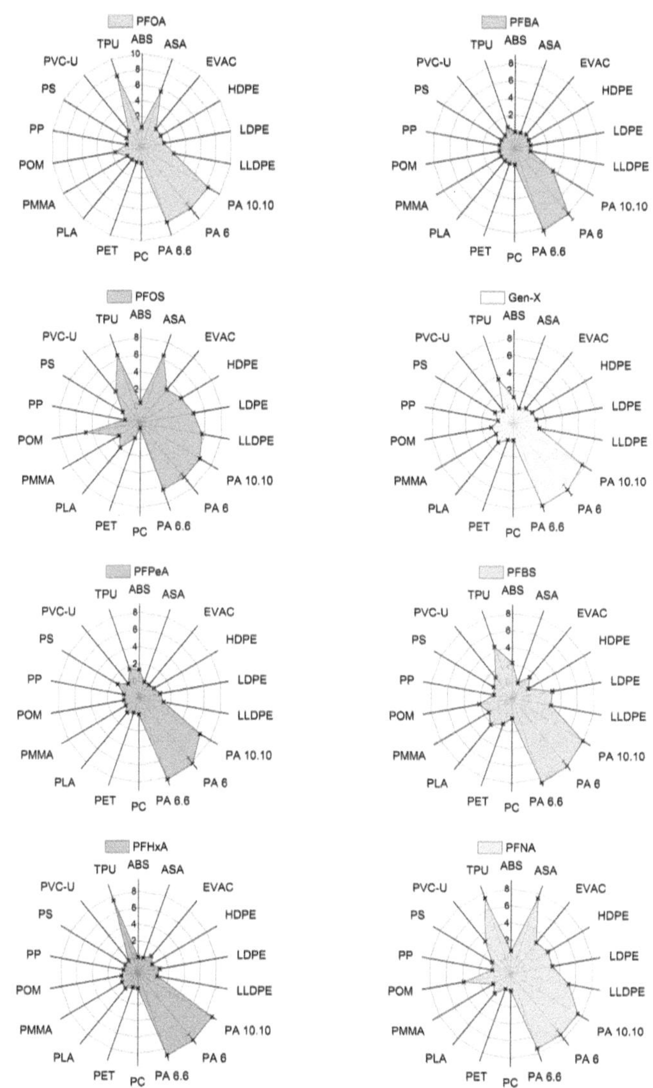

Fig. 3.8 Radial plots displaying uptake of each PFAS in mg per g polymer

dipole–dipole interactions with both the carbonyl and amine portions of the polymer's amide linkage.

Another significant observation pertains to PFAS polarity. Compounds with lower polarity, such as PFNA, adsorb more strongly than those with higher polarity, like PFBA. The longer fluorinated backbone in PFNA augments its hydrophobic character, facilitating stronger attraction to polymer surfaces. In contrast, the more polar PFBA remains soluble in the aqueous phase, resulting in lower overall adsorption. Interestingly, the trend for PFAS polarity does not translate directly to polymer polarity. While PFAS adsorption indeed varies with its own polarity, polymer adsorption efficacy appears to hinge more on specific functional groups (for example, carbonyl, amide, hydroxyl) than on a simple measure of polymer polarity. PP and PE, despite being non-polar and in principle more hospitable to hydrophobic compounds, do not offer the hydrogen-bonding or electrostatic interactions present in amide-rich matrices. Consequently, their PFAS uptake remains low.[135]

Although many studies have found stronger adsorption for long-chain PFAS compared to short-chain analogs, the data here do not reveal a pronounced distinction based on chain length.[134, 136] One possible explanation is that the tested polymers contain polar functional groups, such as amide linkages, that enable hydrogen bonding and electrostatic interactions strong enough to overshadow typical chain-length effects. Longer perfluorinated chains often increase hydrophobic interactions, whereas short chains are more water-soluble and typically exhibit lower sorption.[6] In this study, however, the presence of amide or similarly polar functionalities (for example, urethane in TPU) may dominate the overall binding mechanism. Such polar sites can form stable complexes with both carboxylated and sulfonated PFAS head groups, effectively leveling out the influence of hydrophobic chain length. Moreover, variations in polymer crystallinity or surface accessibility might mitigate typical short-chain versus long-chain discrepancies, enabling even shorter PFAS to adsorb at levels comparable to their longer-chain counterparts.[137]

Overall, the trends are observable: lower-polarity PFAS, such as PFNA, PFOS, and PFOA, exhibit stronger adsorption, while polymers bearing polar functional groups, particularly amide linkages, display the highest adsorption rates. These findings underscore the crucial influence of chemical functionality over bulk polymer polarity and highlight the importance of assessing both hydrophobic and hydrogen-bonding interactions when evaluating PFAS-polymer systems.

3.2.3 Impact of pH on PFAS-Polymer Interactions

Based on their previously observed uptake behaviors and diverse polymer structures, ASA, LDPE, PA 6, PVC, and TPU were selected for the analysis of pH impact. The bar chart in Fig. 3.9 reveals a pronounced effect of pH on PFAS uptake. A higher pH of 10 consistently yields the lowest adsorption values, especially for polymers containing amide or carbonyl functionalities, such as TPU, ASA, and PA 6. This observation aligns with findings by Appleman *et al.*[132] and Du *et al.*[133], which demonstrated that polar functional groups on polymers play a significant role in adsorption under varying pH conditions.

Among the selected polymers, LDPE (low-density polyethylene) and PVC (polyvinyl chloride) essentially lack ionizable groups and thus do not have meaningful pK_a or isoelectric points.[138, 139] Their backbones consist predominantly of nonpolar C–H bonds (LDPE) or C–H/C–Cl bonds (PVC), so their adsorption mechanisms rely primarily on hydrophobic interactions. As a result, the changes in PFAS uptake with increasing pH are comparatively minimal for these two materials.

By contrast, ASA (acrylonitrile-styrene-acrylate) contains polar nitrile and ester groups but typically does not feature free acid sites. The acrylate portion is generally fully esterified, meaning ASA does not exhibit a bulk pK_a or isoelectric point.[91, 138] Nevertheless, ester and nitrile functionalities can engage in dipole-type interactions, and at higher pH, even limited hydrolysis or amine protonation sites could be influenced, contributing to reduced adsorption as observed in Fig. 3.9.

PA 6 (polyamide 6) shows especially striking behavior, as its adsorption of PFAS drops substantially between pH 4 and pH 10. Although PA 6 does not have a single well-defined polymer-wide pK_a, trace amine/carboxyl end groups can exhibit an

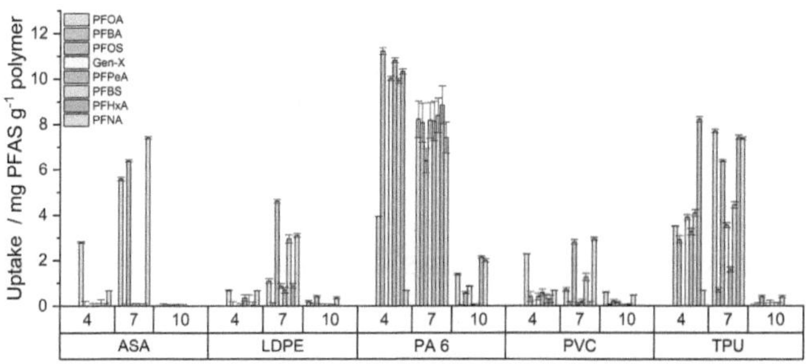

Fig. 3.9 pH-dependent uptake of PFAS in mg per g of polymer, for ASA, LDPE, PA 6, PVC and TPU

so called apparent isoelectric point around pH 3–4.[140] At lower pH, these amine or amide groups undergo protonation, forming positively charged sites that attract negatively charged PFAS molecules through electrostatic interactions, consistent with Goss, who emphasized the role of hydrogen bonding and electrostatic forces in PFAS-polymer interactions.[135]

Similarly, TPU (thermoplastic polyurethane) contains urethane (and sometimes urea) linkages, which have very weakly acidic NH groups (estimated $pK_a \gg 14$), thus lacking a standard pK_a or isoelectric point.[141] Nonetheless, TPU can form protonated sites at low pH, particularly if there are secondary or tertiary amines present, making it effective in binding PFAS via electrostatic attractions. When the pH is raised to 10, fewer protonated groups remain, reducing these interactions.

Another mechanism, though less likely given the low pK_a of PFAS, is the partial protonation of PFAS under acidic conditions, which could promote hydrogen bonding with the polymer's carbonyl groups. This pathway, highlighted by Higgins et al., underscores the potential for increased adsorption due to such interactions.[134] Either mechanism could explain the significant decrease in uptake for amide- or carbonyl-rich polymers when pH is raised to 10, as quaternary amine formation diminishes under basic conditions, reducing electrostatic attractions and leading to lower PFAS adsorption.

Ultimately, these findings reinforce the conclusion that specific chemical functionalities within the polymer backbone, rather than overall polymer polarity alone, govern PFAS adsorption.[137] The pronounced pH sensitivity observed in polymers like PA 6, TPU, and ASA strongly implicates electrostatic interactions and hydrogen bonding as key drivers of PFAS binding in these systems. Appleman et al. noted similar trends, where functionalized adsorbents performed better at low pH due to enhanced binding via these mechanisms.[132] From a practical standpoint, the pH of the surrounding environment significantly alters PFAS removal or partitioning when materials such as polyamides or polyurethanes are involved. Understanding these pH-dependent adsorption properties is crucial for designing effective remediation strategies and predicting the fate of PFAS in various environmental or industrial settings.

3.3 Theoretical Calculations Using a Global Optimization

To examine the nature of substrate interactions with the stationary phase employed during the extraction process, a global optimization was conducted using the neural network potential MACE-OFF23. By analyzing the total energies, presented in

Fig. 3.10, the five most stable conformers for each system were identified. These conformers were then further refined using the GFN2-xTB/GBSA level of theory.

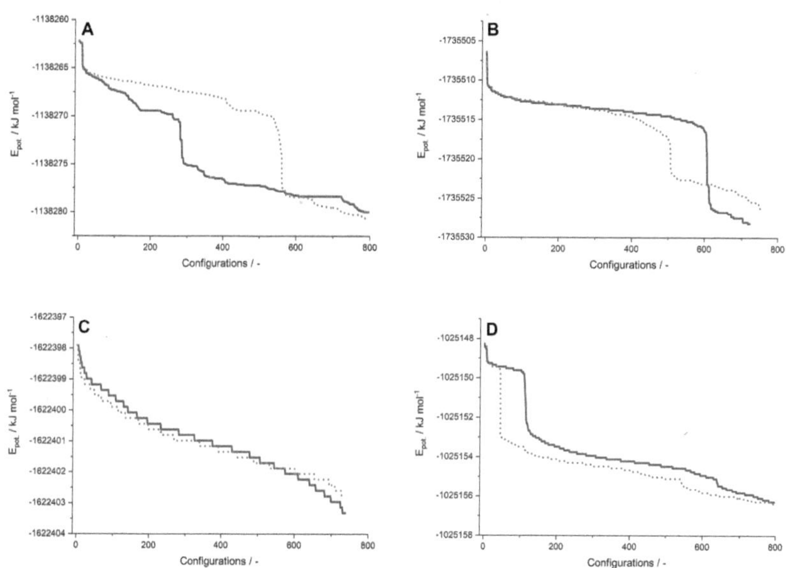

Fig. 3.10 The potential energy of various configurations, ranked in descending order after global optimization using the MACE-OFF23 neural network potential, is shown for A) the PA-PFBA, B) the PA-PFOA, C) the PE-PFBA, and D) the PE-PFOA model systems. For each system, a second optimization (indicated by the dotted line) was performed by mirroring the substrate across the xy-plane, equivalent to a 180° rotation relative to the polymer

The minimum configurations of the PFOA and PFBA substrate molecules with the PE and PA model systems are depicted in Fig. 3.11. The corresponding interactions energies determined using Equation (2.3) were found as -32 and $-39\,\mathrm{kJ\,mol^{-1}}$ in case of the PE-PFBA and PE-PFOA systems respectively, indicating a weak interaction between the PE model system and the substrate molecules. In contrast, values of -126 and $-117\,\mathrm{kJ\,mol^{-1}}$ were found for the PA-PFBA and PA-PFOA systems, which are four and three times more negative compared to the corresponding PE case (Table 3.2). These results provide direct evidence, that the PA substrate is more suited in the extraction of PFAS compared to PE.

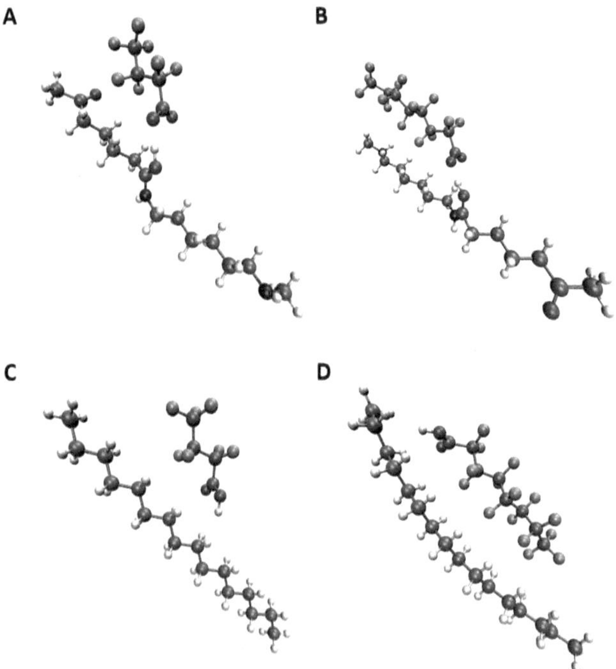

Fig. 3.11 Minimum energy configurations determined for A) PA-PFBA, B) PA-PFOA, C) PE-PFBA, and D) PE-PFOA model systems. These were identified through global optimization using the MACE-OFF23 neural network potential, followed by re-optimization of the five lowest-energy conformers at the GFN2-xTB/GBSA level of theory

Table 3.2 Overview of the interaction energies (E_{Int}) of PFAS with different polymers. The table lists the polymer type, the specific PFAS compound, and the calculated interaction energy calculated from Equation (2.3) in kJ mol^{-1}

Polymer	PFAS	$E_{\text{Int}} / \text{kJ mol}^{-1}$
PE	PFBA	−32
PE	PFOA	−39
PA 6.6	PFBA	−126
PA 6.6	PFOA	−117

3.4 Comparison of Experimental and Theoretical Results

Experimental results reveal that PA consistently exhibit notably higher adsorption of both short- and long-chain PFAS, whereas PE shows comparatively modest uptake. Theoretical calculations using a global optimization approach with the MACE-OFF23 neural network potential (NNP) corroborate these observations by indicating stronger and more energetically favorable interactions (-126 and $-117\,\mathrm{kJ\,mol^{-1}}$) between polyamide and PFBA/PFOA, compared to those with polyethylene (-39 and $-32\,\mathrm{kJ\,mol^{-1}}$). These calculations highlight the role of hydrogen bonding and electrostatic interactions facilitated by amide linkages, aligning with the experimentally observed dominance of amide-rich polymers in adsorbing PFAS.

The agreement between computational and experimental outcomes underscores the importance of specific functional groups in driving PFAS adsorption. While PE's purely hydrocarbon structure offers fewer sites for hydrogen bonding, PA's amide functionality forms robust interactions with the carboxyl or sulfonic acid head groups of PFAS. This synergy between empirical data and molecular modeling provides confidence that the high PFAS adsorption measured for PAs arises chiefly from strong non-covalent forces, rather than from simple hydrophobic interactions alone.

Overall, the theoretical and experimental findings converge on the same conclusion: polymers bearing polar functional groups, particularly amide linkages, exhibit superior adsorption capacities for PFAS. The computational insights further clarify why even short-chain PFAS display strong affinities for these polymers, which is a crucial consideration for optimizing PFAS remediation materials and predicting their behavior in diverse environments.

Conclusion and Outlook $\qquad$ 4

The preliminary experiments underscore the significant impact of vial and filter materials on PFAS recovery. Glass vials caused substantial analyte loss, whereas Waters polypropylene (PP) vials preserved target concentrations accurately. Similarly, cellulose acetate (CA) filters demonstrated consistently minimal PFAS adsorption, making them an optimal choice for sample preparation. Adopting these materials ensures more reliable, reproducible measurements, thereby enhancing the validity of subsequent PFAS adsorption studies and serving as a guideline for selecting proper laboratory consumables in related analyses.

Furthermore, this work provides valuable insights into the complex interplay between per- and polyfluoroalkyl substances (PFAS) and synthetic microplastics (MPs). The experiments confirm that PFAS adsorption strongly depends on polymer composition, interaction time, pH, and PFAS polarity. Polyamides (PAs) reach near-complete adsorption within 24 hours, most likely due to their amide linkages, whereas PET and PP exhibit more gradual or fluctuating uptake driven by sorption-desorption processes. LDPE shows a rapid initial binding followed by partial desorption at longer times. Notably, high pH levels significantly lowers PFAS uptake in amide-rich polymers, highlighting the importance of electrostatic interactions and hydrogen bonding under acidic or neutral conditions.

Theoretical calculations underscore these findings, revealing substantially higher binding energies for PA than for PE. By pinpointing hydrogen bonding as a key mechanism, computational results offer deeper insight into the superior adsorption mechanisms of amide-based polymers. These observations underscore how polymer functionality, PFAS polarity, and environmental conditions collectively shape adsorption outcomes.

While this study provides a critical foundation, further experiments have to be conducted. Investigations into salinity, competing ions, and the extended stability of sorbed PFAS under real-world conditions will help make strategies for PFAS

K. Huter, *Unraveling PFAS-Microplastic Interactions*, BestMasters, https://doi.org/10.1007/978-3-658-49859-7_4

containment and removal. Further, thorough testing of analytical workflows remains crucial, as materials used in laboratory equipment and prolonged storage times can drastically affect measurement results. Additional computational modeling, combined with variations in PFAS chain lengths, concentrations, or pH, could reveal finer details of PFAS adsorption pathways, guiding the development of more selective and robust remediation technologies.

Bibliography

[1]   R. C. Buck, J. Franklin, U. Berger, J. M. Conder, I. T. Cousins, P. de Voogt, A. A. Jensen, K. Kannan, S. A. Mabury, S. P. J. van Leeuwen, *Integr. Environ. Assess. Manag.* **2011**, *7*, 513–541.

[2]   M. Kirk, K. Smurthwaite, J. Braunig, S. Trevenar, C. D'Este, R. Lucas, A. Lal, R. Korda, A. Clements, J. Mueller, B. Armstrong, The PFAS Health Study: Systematic Literature Review, en-AU, **2018**.

[3]   Z. Abunada, M. Y. D. Alazaiza, M. J. K. Bashir, *Water* **2020**, *12*, 3590.

[4]   *Contaminants in our water, Identification & remediation methods*, (Eds.: S. Ahuja, B. G. Loganathan), American Chemical Society, Washington, **2020**, pp. 295.

[5]   B. C. Crone, T. F. Speth, D. G. Wahman, S. J. Smith, G. Abulikemu, E. J. Kleiner, J. G. Pressman, *Crit. Rev. Environ. Sci. Technol.* **2019**, *49*, 2359–2396.

[6]   E. Kissa, *Fluorinated surfactants and repellents*, 2nd ed. rev. and expanded., Marcel Dekker, New York, **2001**, pp. 615.

[7]   Y. Zushi, J. N. Hogarh, S. Masunaga, *Clean. Technol. Envir.* **2012**, *14*, 9–20.

[8]   E. Panieri, K. Baralic, D. Djukic-Cosic, A. Buha Djordjevic, L. Saso, *Toxics* **2022**, *10*.

[9]   D. O'Hagan, *Chem. Soc. Rev.* **2008**, *37*, 308–319.

[10] V. Boiteux, X. Dauchy, C. Rosin, J.-F. Munoz, *Arch. Environ. Contam. Toxicol.* **2012**, *63*, 1–12.

[11] J. Glüge, M. Scheringer, I. T. Cousins, J. C. DeWitt, G. Goldenman, D. Herzke, R. Lohmann, C. A. Ng, X. Trier, Z. Wang, *Environ. Sci.: Process. Impacts* **2020**, *22*, 2345–2373.

[12] L. G. T. Gaines, *Am. J. Ind. Med.* **2023**, *66*, 353–378.

[13] E. Gagliano, M. Sgroi, P. P. Falciglia, F. G. A. Vagliasindi, P. Roccaro, *Water Res.* **2020**, *171*, 115381.

[14] Di Du, Y. Lu, Y. Zhou, Q. Li, M. Zhang, G. Han, H. Cui, E. Jeppesen, *J. Hazard. Mater.* **2021**, *405*, 124681.

[15] S. Brendel, É. Fetter, C. Staude, L. Vierke, A. Biegel-Engler, *Environ. Sci. Eur.* **2018**, *30*, 9.

[16] Z. Wang, J. M. Boucher, M. Scheringer, I. T. Cousins, K. Hungerbühler, *Environ. Sci. Technol.* **2017**, *51*, 4482–4493.

[17] *Summary report on the new comprehensive global database of Per- and Polyfluoroalkyl Substances (PFASs)*, OECD, **2018**.

[18] A. M. Calafat, L.-Y. Wong, Z. Kuklenyik, J. A. Reidy, L. L. Needham, *Environ. Health Perspect.* **2007**, *115*, 1596–1602.

[19] A. M. Becker, S. Gerstmann, H. Frank, *Chemosphere* **2008**, *72*, 115–121.

[20] A. Pistocchi, R. Loos, *Environ. Sci. Technol.* **2009**, *43*, 9237–9244.

[21] K. Prevedouros, I. T. Cousins, R. C. Buck, S. H. Korzeniowski, *Environ. Sci. Technol.* **2006**, *40*, 32–44.

[22] J. M. Armitage, M. MacLeod, I. T. Cousins, *Environ. Sci. Technol.* **2009**, *43*, 1134–1140.

[23] European Commission—Have your say, Chemical pollutants—restrictions on perfluorooctanoic acid (PFOA), **2024**, https://ec.europa.eu/info/law/betterregulation/have-your-say/initiatives/12057-Chemical-pollutantsrestrictions-on-perfluorooctanoic-acid-PFOA-_en (visited on 06/10/2024).

[24] BUND Naturschutz, Eine kurze PFOA Geschichte, https://altoetting.bund-naturschutz.de/fileadmin/kreisgruppen/altoetting/pdf/Themen/Wasser/Trinkwasser_PFOA/PFOAkurzeGeschichte_01.pdf (visited on 06/10/2024).

[25] US EPA, Fact Sheet: 2010/2015 PFOA Stewardship Program—US EPA, **2016**, https://www.epa.gov/assessing-and-managing-chemicals-under-tsca/factsheet-20102015-pfoa-stewardship-program (visited on 06/10/2024).

[26] A. Shankar, J. Xiao, A. Ducatman, *Arch. Intern. Med.* **2012**, *172*, 1397–1403.

[27] S. J. Frisbee, A. P. Brooks, A. Maher, P. Flensborg, S. Arnold, T. Fletcher, K. Steenland, A. Shankar, S. S. Knox, C. Pollard, J. A. Halverson, V. M. Vieira, C. Jin, K. M. Leyden, A. M. Ducatman, *Environ. Health Perspect.* **2009**, *117*, 1873–1882.

[28] V. Barry, A. Winquist, K. Steenland, *Environ. Health Perspect.* **2013**, *121*, 1313–1318.

[29] V. M. Vieira, K. Hoffman, H.-M. Shin, J. M. Weinberg, T. F. Webster, T. Fletcher, *Environ. Health Perspect.* **2013**, *121*, 318–323.

[30] A. Cordner, V. Y. de La Rosa, L. A. Schaider, R. A. Rudel, L. Richter, P. Brown, *J. Expo. Sci. Environ. Epidemiol.* **2019**, *29*, 157–171.

[31] Minnesota Department of Health, Environmental Health Division, Health Risk Assessment Unit, PFBA Info Sheet April, **2022**.

[32] H. Su, Y. Shi, Y. Lu, P. Wang, M. Zhang, A. Sweetman, K. Jones, A. Johnson, *Environ. Int.* **2017**, *101*, 1–6.

[33] B. M. Sharma, G. K. Bharat, S. Tayal, T. Larssen, J. Bečanová, P. Karásková, P. G. Whitehead, M. N. Futter, D. Butterfield, L. Nizzetto, *Environ. Pollut.* **2016**, *208*, 704–713.

[34] U.S. EPA, IRIS Toxicological Review of Perfluorobutanoic Acid (PFBA) and Related Salts (Final Report, 2022), U.S. Environmental Protection Agency, Washington, DC, EPA/635/R-22/277F, **2021**, https://cfpub.epa.gov/ncea/iris_drafts/recordisplay.cfm?deid=356425 (visited on 06/11/2024).

[35] K. Abraham, A. H. El-Khatib, T. Schwerdtle, B. H. Monien, *Int. J. Hyg. Environ.* **2021**, *237*, 113830.

[36] L. M. Weatherly, H. L. Shane, E. Lukomska, R. Baur, S. E. Anderson, *Food Chem. Toxicol.* **2021**, *156*, 112528.

[37] S. Liu, R. Yang, N. Yin, F. Faiola, *J. Environ. Sci.* **2020**, *88*, 187–199.

[38] K. P. Das, B. E. Grey, R. D. Zehr, C. R. Wood, J. L. Butenhoff, S.-C. Chang, D. J. Ehresman, Y.-M. Tan, C. Lau, *Toxicol. Sci.* **2008**, *105*, 173–181.

[39] R. Renner, *Environ. Sci. Technol.* **2000**, *34*, 371A–3A.

[40] B. Ameduri, *Chemistry* **2018**, *24*, 18830–18841.

[41] D. Z. Wang, G. Goldenman, T. Tugran, A. McNeil, M. Jones, Per- and polyfluoroalkylether substances: identity, production and use, **2020**.

[42] Organization for Economic Cooperation and Development, Risk reduction approaches for PFAS. A cross-country analysis, Environment, Health and Safety Publications. Series on Risk Management No. 29. Paris, France: OECD, **2015**, https://www.oecd.org/chemicalsafety/riskmanagement/Risk_Reduction_Approaches 06/11/2024).

[43] A. Blum, S. A. Balan, M. Scheringer, X. Trier, G. Goldenman, I. T. Cousins, M. Diamond, T. Fletcher, C. Higgins, A. E. Lindeman, G. Peaslee, P. de Voogt, Z. Wang, R. Weber, *Environ. Health Perspect.* **2015**, *123*, A107–11.

[44] T. Fletcher, T. S. Galloway, D. Melzer, P. Holcroft, R. Cipelli, L. C. Pilling, D. Mondal, M. Luster, L. W. Harries, *Environ. Int.* **2013**, *57–58*, 2–10.

[45] X. He, Y. Liu, B. Xu, L. Gu, W. Tang, *Sci. Total Environ.* **2018**, *625*, 566–574.

[46] N. Johansson, A. Fredriksson, P. Eriksson, *Neurotoxicology* **2008**, *29*, 160–169.

[47] Y. Yue, S. Li, Z. Qian, R. F. Pereira, J. Lee, J. J. Doherty, Z. Zhang, Y. Peng, J. M. Clark, A. R. Timme-Laragy, Y. Park, *Food Chem. Toxicol.* **2020**, *145*, 111695.

[48] L. Zheng, G.-H. Dong, Y.-H. Jin, Q.-C. He, *Arch. Toxicol.* **2009**, *83*, 679–689.

[49] D. Melzer, N. Rice, M. H. Depledge, W. E. Henley, T. S. Galloway, *Environ. Health Perspect.* **2010**, *118*, 686–692.

[50] G. Du, J. Hu, Z. Huang, M. Yu, C. Lu, X. Wang, Di Wu, *Ecotoxicol. Environ. Saf.* **2019**, *167*, 412–421.

[51] A. Florentin, T. Deblonde, N. Diguio, A. Hautemaniere, P. Hartemann, *Int. J. Hyg. Environ.* **2011**, *214*, 493–499.

[52] Z. R. Hopkins, M. Sun, J. C. DeWitt, D. R. Knappe, *J. Am. Water Works Assoc.* **2018**, *110*, 13–28.

[53] M. Sun, E. Arevalo, M. Strynar, A. Lindstrom, M. Richardson, B. Kearns, A. Pickett, C. Smith, D. R. U. Knappe, *Environ. Sci. Technol. Lett.* **2016**, *3*, 415–419.

[54] F. Heydebreck, J. Tang, Z. Xie, R. Ebinghaus, *Environ. Sci. Technol.* **2015**, *49*, 8386–8395.

[55] W. A. Gebbink, L. van Asseldonk, S. P. J. van Leeuwen, *Environ. Sci. Technol.* **2017**, *51*, 11057–11065.

[56] J. M. Caverly Rae, L. Craig, T. W. Slone, S. R. Frame, L. W. Buxton, G. L. Kennedy, *Toxicol. Rep.* **2015**, *2*, 939–949.

[57] S. Lerner, New Teflon Toxin Causes Cancer in Lab Animals, **2016**, https://theintercept.com/2016/03/03/new-teflon-toxin-causes-cancer-inlab-animals/ (visited on 06/12/2024).

[58] Sigma-Aldrich, Perfluoropentanoic acid 97 2706-90-3, **2024**, https://www.sigmaaldrich.com/AT/en/product/aldrich/396575 (visited on 06/12/2024).

[59] M.-K. Kim, T. Kim, T.-K. Kim, S.-W. Joo, K.-D. Zoh, *Sep. Purif. Technol.* **2020**, *247*, 116911.

[60] J. Bangma, J. McCord, N. Giffard, K. Buckman, J. Petali, C. Chen, D. Amparo, B. Turpin, G. Morrison, M. Strynar, *Chemosphere* **2023**, *315*, 137722.

[61] D.-H. Kim, M.-Y. Lee, J.-E. Oh, *Environ. Pollut.* **2014**, *192*, 171–178.

[62] ECHA, Substance Information—ECHA, **2024**, https://echa.europa.eu/de/substance-information/-/substanceinfo/100.282.094 (visited on 06/12/2024).

[63] L. Jiang, Y. Hong, G. Xie, J. Zhang, H. Zhang, Z. Cai, *Sci. Total Environ.* **2021**, *790*, 148160.

[64] K. M. Annunziato, C. E. Jantzen, M. C. Gronske, K. R. Cooper, *Aquat. Toxicol.* **2019**, *208*, 126–137.

[65] J. L. Butenhoff, S.-C. Chang, D. J. Ehresman, R. G. York, *Reprod. Toxicol.* **2009**, *27*, 331–341.

[66] X. Guo, S. Zhang, X. Liu, S. Lu, Q. Wu, P. Xie, *Ecotoxicol. Environ. Saf.* **2021**, *225*, 112733.

[67] J. E. Klaunig, M. Shinohara, H. Iwai, C. P. Chengelis, J. B. Kirkpatrick, Z. Wang, R. H. Bruner, *Toxicol. Pathol.* **2015**, *43*, 209–220.

[68] Y. Liu, M. M. Bahar, S. V. A. C. Samarasinghe, F. Qi, S. Carles, W. R. Richmond, Z. Dong, R. Naidu, *J. Hazard. Mater.* **2022**, *439*, 129667.

[69] A. Presentato, S. Lampis, A. Vantini, F. Manea, F. Daprà, S. Zuccoli, G. Vallini, *Microorganisms* **2020**, *8*, 92.

[70] X. Fang, L. Zhang, Y. Feng, Y. Zhao, J. Dai, *Toxicol. Sci.* **2008**, *105*, 312–321.

[71] C. C. Bach, B. H. Bech, E. A. Nohr, J. Olsen, N. B. Matthiesen, R. Bossi, N. Uldbjerg, E. C. Bonefeld-Jørgensen, T. B. Henriksen, *Environ. Res.* **2015**, *142*, 535–541.

[72] M. Houde, R. S. Wells, P. A. Fair, G. D. Bossart, A. A. Hohn, T. K. Rowles, J. C. Sweeney, K. R. Solomon, D. C. G. Muir, *Environ. Sci. Technol.* **2005**, *39*, 6591–6598.

[73] *Polystyrene: Synthesis, Characteristics and Applications*, (Ed.: L. Cole), Nova Science Publishers, **2014**.

[74] R. C. Thompson, Y. Olsen, R. P. Mitchell, A. Davis, S. J. Rowland, A. W. G. John, D. McGonigle, A. E. Russell, *Science* **2004**, *304*, 838.

[75] Z. Akdogan, B. Guven, *Environ. Pollut.* **2019**, *254*, 113011.

[76] M. C. Rillig, A. Lehmann, *Science* **2020**, *368*, 1430–1431.

[77] D. M. Mitrano, W. Wohlleben, *Nat. Commun.* **2020**, *11*, 5324.

[78] S. Ghosh, J. K. Sinha, S. Ghosh, K. Vashisth, S. Han, R. Bhaskar, *Sustainability* **2023**, *15*, 10821.

[79] A. A. de Souza Machado, W. Kloas, C. Zarfl, S. Hempel, M. C. Rillig, *Glob. Change Biol.* **2018**, *24*, 1405–1416.

[80] S. Lambert, C. Scherer, M. Wagner, *Integr. Environ. Assess. Manag.* **2017**, *13*, 470–475.

[81] G. Caruso, *Mar. Pollut. Bull.* **2019**, *146*, 921–924.

[82] P. Schwabl, S. Köppel, P. Königshofer, T. Bucsics, M. Trauner, T. Reiberger, B. Liebmann, *Ann. Intern. Med.* **2019**, *171*, 453–457.

[83] R. Qi, D. L. Jones, Z. Li, Q. Liu, C. Yan, *Sci. Total Environ.* **2020**, *703*, 134722.

[84] K. D. Cox, G. A. Covernton, H. L. Davies, J. F. Dower, F. Juanes, S. E. Dudas, *Environ. Sci. Technol.* **2019**, *53*, 7068–7074.

[85] L. G. A. Barboza, L. R. Vieira, V. Branco, N. Figueiredo, F. Carvalho, C. Carvalho, L. Guilhermino, *Aquat. Toxicol.* **2018**, *195*, 49–57.

[86] J. C. Prata, J. P. Da Costa, I. Lopes, A. C. Duarte, T. Rocha-Santos, *Sci. Total Environ.* **2020**, *702*, 134455.

[87] S. L. Wright, F. J. Kelly, *Environ. Sci. Technol.* **2017**, *51*, 6634–6647.

[88] A. D. Vethaak, J. Legler, *Science* **2021**, *371*, 672–674.

[89] Manish, D. Gurjar, S. Sharma, Akash, M. Sarkar, *Mater. Today* **2018**, *5*, 28296–28304.

[90] J. Moore, *Compos.* **1973**, *4*, 118–130.

[91] J. Scheirs, D. B. Priddy, *Modern Styrenic Polymers: Polystyrenes and Styrenic Copolymers*, Wiley, **2003**.

[92] A. M. Henderson, *IEEE Electr. Insul. Mag.* **1993**, *9*, 30–38.

[93] S. Ronca in *Brydson's Plastics Materials*, Elsevier, **2017**, pp. 247–278.

[94] Engr. Mohammed Zillane Patwary, Physical and Chemical Properties Of Nylon 6, (Ed.: Textile Fashion Study), Textile Fashion Study, **2012**, https://textilefashionstudy. com/polyamide-fiber-physical-and-chemicalproperties-of-nylon-6/ (visited on 01/07/2025).

[95] Reichelt Chemietechnik GmbH + Co., Polyamid 6.6, https://www.rct-online.de/de/ RctGlossar/detail/id/7 (visited on 01/07/2025).

[96] M. Winnacker, B. Rieger, *Macromol. Rapid Commun.* **2016**, *37*, 1391–1413.

[97] J. Pagacz, K. N. Raftopoulos, A. Leszczyńska, K. Pielichowski, *J. Therm. Anal. Calorim.* **2016**, *123*, 1225–1237.

[98] A. Davis, J. H. Golden, *J. Macromol. Sci. Rev. C* **1969**, *3*, 49–68.

[99] K. Ravindranath, R. A. Mashelkar, *Chem. Eng. Sci.* **1986**, *41*, 2197–2214.

[100] X. Pang, X. Zhuang, Z. Tang, X. Chen, *Biotechnol. J.* **2010**, *5*, 1125–1136.

[101] U. Ali, K. J. B. A. Karim, N. A. Buang, *Polym. Rev.* **2015**, *55*, 678–705.

[102] S. Lüftl, V. P.M., S. Chandran, *Polyoxymethylene Handbook*, Wiley, **2014**.

[103] P. Lieberzeit, D. Bekchanov, M. Mukhamediev, *Polym. Adv. Technol.* **2022**, *33*, 1809–1820.

[104] V. Busico, R. Cipullo, *Prog. Polym. Sci.* **2001**, *26*, 443–533.

[105] J. Datta, P. Kasprzyk, *Polym. Eng. Sci.* **2018**, *58*.

[106] A. A. Horton, D. K. A. Barnes, *Sci. Total Environ.* **2020**, *738*, 140349.

[107] Ivar do Sul, Juliana A., M. F. Costa, *Environ. Pollut.* **2014**, *185*, 352–364.

[108] J. R. Larsen, J. Provencher, E. Farmen, *Litter and Microplastics: Environmental monitoring in the Arctic*, **2021**.

[109] M. C. Rillig, *Environ. Sci. Technol.* **2012**, *46*, 6453–6454.

[110] M. Ateia, T. Zheng, S. Calace, N. Tharayil, S. Pilla, T. Karanfil, *Sci. Total Environ.* **2020**, *720*, 137634.

[111] Y. Shi, H. Almuhtaram, R. C. Andrews, *Polymers* **2023**, *15*, 3676.

[112] J. W. Scott, K. G. Gunderson, L. A. Green, R. R. Rediske, A. D. Steinman, *Toxics* **2021**, *9*, 106.

[113] P. J. Brahana, A. Al Harraq, L. E. Saab, R. Roberg, K. T. Valsaraj, B. Bharti, *Environ. Sci.: Process. Impacts* **2023**, *25*, 1519–1531.

[114] F. Wang, K. M. Shih, X. Y. Li, *Chemosphere* **2015**, *119*, 841–847.

[115] S. Lath, E. R. Knight, D. A. Navarro, R. S. Kookana, M. J. McLaughlin, *Chemosphere* **2019**, *222*, 671–678.

[116] B. Barhoumi, S. G. Sander, I. Tolosa, *Environ. Res.* **2022**, *206*, 112595.

[117] S. P. Lenka, M. Kah, L. P. Padhye, *ACS ES&T Water* **2023**, *3*, 2700–2706.

[118] Z. Ning, S. Zhou, Y. Yang, P. Li, Z. Zhao, W. Zhang, L. Lu, N. Ren, *Water Environ. Res.* **2024**, *96*, e11080.

[119] B. Wen, S.-R. Jin, Z.-Z. Chen, J.-Z. Gao, Y.-N. Liu, J.-H. Liu, X.-S. Feng, *Environ. Pollut.* **2018**, *243*, 462–471.

[120] R. Bakhshoodeh, R. M. Santos, *RSC Adv.* **2022**, *12*, 4973–4987.

[121] D. C. Harris, *Lehrbuch der Quantitativen Analyse*, 8. Aufl. 2014, (Eds.: G. Werner, T. Werner), Springer Berlin Heidelberg, Berlin, Heidelberg, **2014**, pp. 977.

[122] E. de Hoffmann, V. Stroobant, *Mass spectrometry, Principles and applications*, 2. ed., Reprint, Wiley, Chichester and Weinheim, **2006**, pp. 407.

[123] J. Lee, H. T. Gan, S. M. A. Latiff, C. Chuah, W. Y. Lee, Y.-S. Yang, B. Loo, S. K. Ng, P. Gagnon, *J. Chromatogr. A* **2012**, *1270*, 162–170.

[124] D. P. Kovács, J. H. Moore, N. J. Browning, I. Batatia, J. T. Horton, V. Kapil, W. C. Witt, I.-B. Magdău, D. J. Cole, G. Csányi, MACE-OFF23: Transferable Machine Learning Force Fields for Organic Molecules, **2023**.

[125] D. J. Wales, J. P. K. Doye, *J. Phys. Chem. A* **1997**, *101*, 5111–5116.

[126] C. Bannwarth, E. Caldeweyher, S. Ehlert, A. Hansen, P. Pracht, J. Seibert, S. Spicher, S. Grimme, *Wiley Interdiscip. Rev. Comput. Mol. Sci.* **2021**, *11*.

[127] C. Bannwarth, S. Ehlert, S. Grimme, *J. Chem. Theory Comput.* **2019**, *15*, 1652–1671.

[128] S. Grimme, C. Bannwarth, P. Shushkov, *J. Chem. Theory Comput.* **2017**, *13*, 1989–2009.

[129] B. Hourahine, B. Aradi, V. Blum, F. Bonafé, A. Buccheri, C. Camacho, C. Cevallos, M. Y. Deshaye, T. Dumitrică, A. Dominguez, S. Ehlert, M. Elstner, T. van der Heide, J. Hermann, S. Irle, J. J. Kranz, C. Köhler, T. Kowalczyk, T. Kubař, I. S. Lee, V. Lutsker, R. J. Maurer, S. K. Min, I. Mitchell, C. Negre, T. A. Niehaus, A. M. N. Niklasson, A. J. Page, A. Pecchia, G. Penazzi, M. P. Persson, J. Řezáč, C. G. Sánchez, M. Sternberg, M. Stöhr, F. Stuckenberg, A. Tkatchenko, V. W.-Z. Yu, T. Frauenheim, *J. Chem. Phys.* **2020**, *152*, 124101.

[130] Grimme-Lab, Parameter files for the GBSA implicit solvation model, **2020**, https://github.com/grimme-lab/gbsa-parameters

[131] M. B. Woudneh, B. Chandramouli, C. Hamilton, R. Grace, *Environ. Sci. Technol.* **2019**, *53*, 12576–12585.

[132] T. D. Appleman, C. P. Higgins, O. Quiñones, B. J. Vanderford, C. Kolstad, J. C. Zeigler-Holady, E. R. V. Dickenson, *Water Res.* **2014**, *51*, 246–255.

[133] Z. Du, S. Deng, Y. Bei, Q. Huang, B. Wang, J. Huang, G. Yu, *J. Hazard. Mater.* **2014**, *274*, 443–454.

[134] C. P. Higgins, R. G. Luthy, *Environ. Sci. Technol.* **2006**, *40*, 7251–7256.

[135] K.-U. Goss, *Environ. Sci. Technol.* **2008**, *42*, 5032.

[136] L. Ahrens, *J. Environ. Monit.* **2011**, *13*, 20–31.

[137] Z. Wang, J. C. DeWitt, C. P. Higgins, I. T. Cousins, *Environ. Sci. Technol.* **2017**, *51*, 2508–2518.

[138] *Polymer handbook*, 4. ed., (Ed.: J. Brandrup), Wiley, New York and Weinheim, **1999**, pp. 2288.

[139] *Modern plastics handbook*, First edition, (Ed.: C. A. Harper), in collab. with C. A. Harper, A.-M. M. Baker, J. Mead, R. E. Wright, J. M. Margolis, L. Kattas, F. Gastrock, I. Levin, A. Cacciatore, C. M. F. Barry, S. A. Orroth, J. L. Hull, W. R. Lukaszyk, P. Stoughton, P. Kennedy, J. I. Rotheiser, E. M. Petrie, C. P. Izzo, R. Shastri, S. E. Selke, McGraw-Hill Education and McGraw Hill, New York, N.Y., **2000**, pp. 900.

[140] F. W. Billmeyer, *Textbook of polymer science*, 3. ed., Wiley, New York, **1984**, pp. 578.

[141] M. Szycher, *Szycher's handbook of polyurethanes*, 2. ed., CRC Press / Taylor & Francis, Boca Raton, Fla., **2013**, pp. 1126.

MIX
Papier aus verantwortungsvollen Quellen
Paper from responsible sources
FSC® C105338

If you have any concerns about our products,
you can contact us on
ProductSafety@springernature.com

In case Publisher is established outside the EU,
the EU authorized representative is:
Springer Nature Customer Service Center GmbH
Europaplatz 3, 69115 Heidelberg, Germany

Printed by Libri Plureos GmbH
in Hamburg, Germany